Islands

ISLANDS

H. W. Menard

**SCIENTIFIC
AMERICAN
LIBRARY**

A division of HPHLP
New York

This book is number 17 of a series.

Library of Congress Cataloging-in-Publication Data

Menard, Henry W. (Henry William), 1920–86
    Islands.

    Bibliography: p.
    Includes index.
    1. Islands. 2. Island ecology. I. Title.
GB471.M46 1986   551.4′2   86-6573
ISBN 0-7167-5017-1

Printed in the United States of America

Scientific American Library
A division of HPHLP
New York

Distributed by W. H. Freeman and Company,
41 Madison Avenue, New York, New York 10010
and 20 Beaumont Street, Oxford, OX1 2NQ, England.

3 4 5 6 7 8 9 0 KP 5 4 3 2 1 0 8 9 8 7

*Again to Gifford, whose wifely love has sustained me
for thirty-nine years and twenty-five expeditions.
In a world adrift, we have always managed to drift together.*

Contents

Editor's Note

Bill Menard died not long before this book was published. He had watched over its preparation with intense interest, but events overtook him before he could write a preface. Had he done so, he surely would have assumed all responsibility for errors. It is my sad duty to claim a share of that responsibility, and to hope that it is a light one.

A. K.

Oceanic Islands of the World

The locations of islands and groups mentioned in this
book are shown in the maps on the next three pages. In
the past two decades, many of these names have changed,
reflecting the shift of autonomy from distant continental
powers to the islands themselves. Many atlases now in
print still use the older names; newer reference works
include both. On the following pages, most of the names
are the familiar older ones; all of them are the names used
in the book itself.

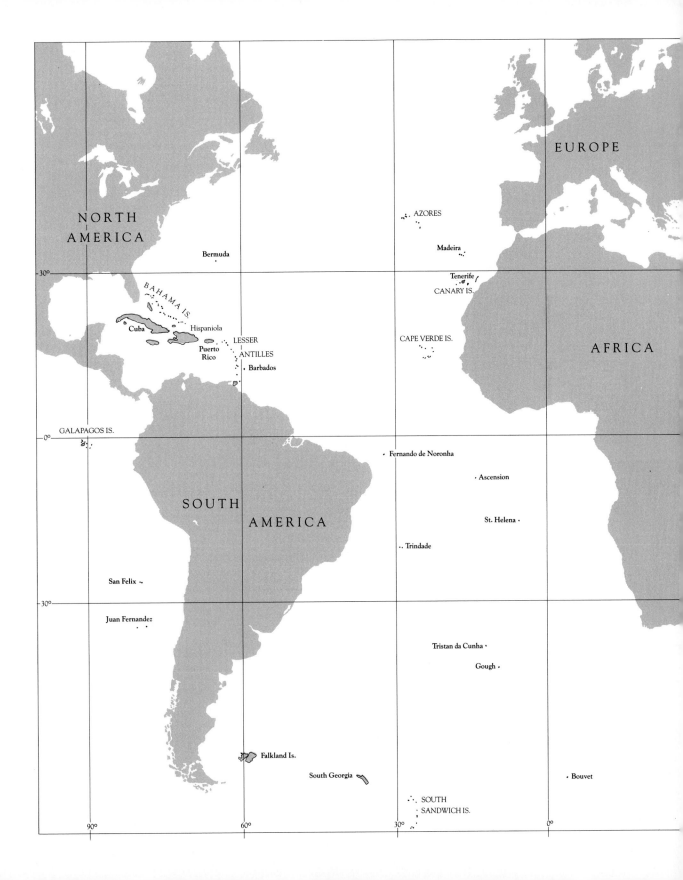

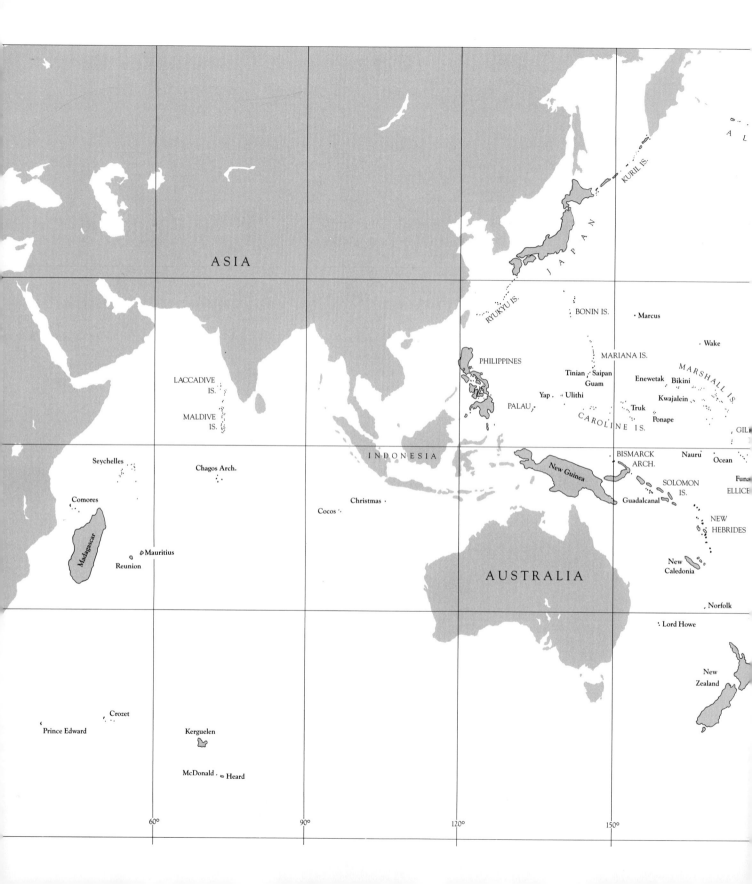

ASIA

KURIL IS.

JAPAN

AL

RYUKYU IS.

BONIN IS.

· Marcus

· Wake

MARIANA IS.

MARSHALL IS.

PHILIPPINES

Tinian · Saipan
Guam

Enewetak   Bikini

Yap · Ulithi

Kwajalein

PALAU ·

Truk ·

Ponape

CAROLINE IS.

GIL

LACCADIVE
IS.

MALDIVE
IS.

INDONESIA

BISMARCK
ARCH.

Nauru   Ocean

Seychelles

Chagos Arch.

New Guinea

SOLOMON
IS.

Funa

ELLICE

Comores

Christmas ·

Guadalcanal

Cocos ·

NEW
HEBRIDES

Madagascar

ᵔ Mauritius

Reunion

AUSTRALIA

New
Caledonia

· Norfolk

·: Lord Howe

New
Zealand

· Crozet

Prince Edward

Kerguelen

McDonald · ᵔ Heard

60°   90°   120°   150°

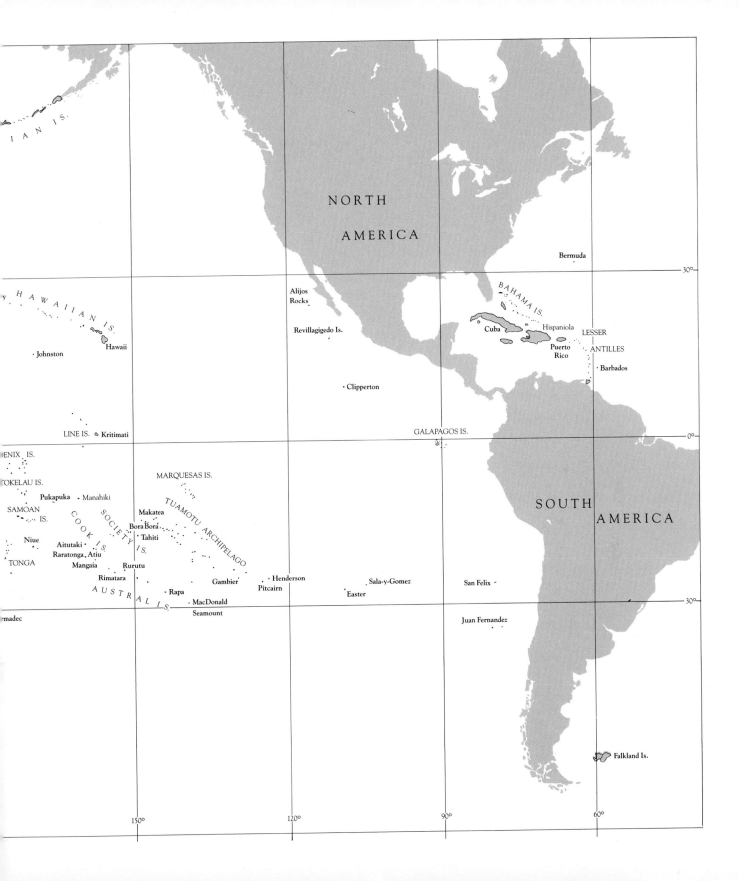

N IS.

IAN IS.

NORTH

AMERICA

Bermuda

30°

HAWAIIAN IS.

Alijos
Rocks

BAHAMA IS.

· Johnston

Revillagigedo Is.

Hawaii

Cuba

Hispaniola

LESSER

Puerto
Rico

ANTILLES

· Barbados

· Clipperton

LINE IS. ◦ Kritimati

GALAPAGOS IS.

0°

ENIX IS.

MARQUESAS IS.

OKELAU IS.

SOUTH

AMERICA

Pukapuka · Manahiki

TUAMOTU ARCHIPELAGO

SAMOAN

Makatea

IS.

COOK IS.

SOCIETY

BoraBora

Niue

· Tahiti

IS.

Aitutaki

Raratonga · Atiu

TONGA

Mangaia · Rurutu

Rimatara

Gambier

· Henderson

· Sala-y-Gomez

San Felix

· Rapa

Pitcairn

Easter

AUSTRAL IS.

· MacDonald

30°

rmadec

Seamount

Juan Fernandez

Falkland Is.

150°

120°

90°

60°

Islands

1

Finding Islands

Charles Darwin was one of the first scientists to learn the advantages of investigating oceanic islands. The scientific method in the laboratory is to isolate a sample of known properties and then observe the effects of systematic changes in pressure, temperature, or some other variable. Nature rarely conducts her experiments under such controlled conditions, but for some studies islands provide her closest approximation to a laboratory. Oceanic islands are small, young, isolated, simple, and subjected to a limited range of environmental factors. Thus, in nature's young, isolated laboratory of the Galapagos Islands, Darwin found the glimmerings of biological evolution. Likewise, the simplicity of several widely separated islands helped him to realize that their differences were largely a consequence of a single geological factor—subsidence.

Consider the continents. They are aggregates of every type of rock produced for billions of years, and most of their history is obscure. Their rocks have been deformed repeatedly, fractured, and warped up and down. They have been eroded and weathered by every type of changing climate, and older rocks are partially buried by thick sedimentary rock derived from them. The whole is obscured by every type of soil and by plants. Across the continents migrate animals and plants in constant flux. One can have little reason to hope that nature has conducted many controlled experiments on the continents. Consider the arsenic that modern chemistry has identified in Napoleon's hair. If he had died in Paris, his poisoners might have been anyone and their motives unknown. However, he died on St. Helena, a small, isolated, volcanic island in the South Atlantic, and all his food came from his British gaolers or his few friends.

The intersection of Polynesian and European cultures painted by William Hodges, who was with Captain Cook at Tahiti.

Maiao Island in the Society Island group—an isolated
peak in a vast sea.

Some islands are merely continents in miniature, and they are difficult to understand for the same reasons. Among these are all the islands
rising from the shallow waters of the continental shelf, islands such as
Ireland and Newfoundland. Likewise, more isolated Japan, New Zealand,
and many other islands have all the geologic characteristics of continents
except size. Even the tiny Seychelles Islands in the Indian Ocean must be
excluded from our story despite their tropic beaches and coconut palms.
They are composed of a granite 700 million years old—both the type and
the age of the rocks show that the Seychelles are a tiny fragment of drifting
continent.

The remaining oceanic islands and their submarine counterparts, seamounts, are remarkably similar. They arise in deep water on normal oceanic crust. All of the thousands of islands, about 20,000 larger seamounts,
and countless smaller ones grew as volcanoes composed of rocks of very
similar types—at least to the nonspecialist. All have come into existence
during the last few percent of the history of the earth. The older ones that
grew high enough to become islands have now sunk beneath the waves.
The only major variable affecting the present appearance of such islands is
that some of them have been in tropic waters and remained there, so they
are now capped with coral reefs in the form of atolls. With such a large and
uniform population of islands, it is relatively easy to isolate the effects of
single variables in nature's experiments. One can compare erosion in the

belt of the trade winds with erosion by polar ice. Likewise, one can compare the number of species that can drift across an oceanic gap of 100 miles with the number that can cross 1000 miles.

THE PROBLEM OF DISCOVERY

The plant and animal life of small isolated islands, like their geology, has intrigued scientists since it was first discovered. How could insect species typical of North America and Asia have reached the Hawaiian Islands? How could an odd creature like the dodo come to be on only one island in the world? Darwin studied the different finches on individual islands in the Galapagos, and Alfred Wallace studied the life of islands in general; between them they produced the theory of biological evolution. No theory ever generated more controversy, but, in more modest ways, almost all ideas about island life have been controversial.

How did plants and animals find and populate oceanic islands? The range of ideas on the subject is remarkable. At one extreme is the idea that most of the biota of islands consists of waifs who drifted there on air or water. At the other extreme is the idea that the islands are peaks of former continents, on which animals walked dry-shod carrying seeds with them. Moreover, it is not just the nonhuman discovery of islands that is controversial. Anthropologists have waxed hot about the Polynesian discovery and occupation of the islands of the central Pacific. It is well to remember that almost every island was successively found and populated by plants, animals, non-Europeans, and Europeans. As in most matters, the less information available, and the narrower the focus, the greater the range of speculation about causes of phenomena. Consequently it seems reasonable to consider what we know least about—the migrations of plants and animals—only after an analysis of the migration of humans. Likewise, it seems only reasonable to consider how Europeans found islands before thinking about how Polynesians did it. It might seem that we need go no further than the history of European voyages to understand how islands are discovered, but, regrettably, even that history is flawed. Imaginative scholarship has worked out when each Pacific island was found by Europeans. (That is the sort of date that appears in history books, the recorded date of the last discovery by any species.) However, data are scanty about the total number of voyages, the efficiency of the discovery effort, and much else that would be useful for generalizations about discovery by other cultures and species. Thus, to obtain an adequate data set, we must go farther afield.

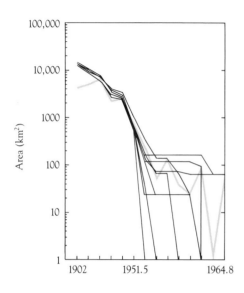

Area (km²)

100,000
10,000
1000
100
10
1
1902   1951.5   1964.8

Comparison of the actual history of exploration for oil in the United States with ten random searches by a computer. The actual rate of discovery per unit effort was no better than a random search. Horizontal units are not equal time periods, but periods of equal exploratory drilling effort. A few times are indicated.

Exploration for Oil

Oceanic islands are small objects that generally occur in clusters separated by vast empty spaces. So are oil fields, and the history of discovery of oil fields is known in great detail, especially in the United States, because of legal requirements for disclosure. The significance of that history was first understood by M. King Hubbert, who developed a measure of effort for oil exploration—the total length of holes drilled with the objective of seeking new oil fields. (Holes that merely expanded known oil fields did not count.) He could estimate the volume of oil discovered each year and compare it with the drilling data. All the oil ultimately found in a field he credited to the year that the field was discovered. Hubbert found that the discoveries per unit effort had declined exponentially with time for more than 80 years. Assuming that history would repeat itself, he could determine the volume of undiscovered oil by simple extrapolation, and in 1967 he predicted a calamitous drop in oil discoveries in the United States. That prediction, properly interpreted, was correct.

In 1975, George Sharman and I applied Hubbert's basic idea to the same problem, namely, how much oil remains to be discovered in the forty-eight contiguous United States, but our analysis can be extended to discovery in general.

We assumed that the chance of discovering an oil field by purely random drilling was simply the area of the oil field divided by the area being explored. For example, the total area of all known giant fields was 23,455 km² and the area of sedimentary basins being explored was 4,700,000 km². Therefore the chance of hitting a giant field with one hole was about 1/200. If most of the oil were in the biggest fields, purely random drilling would discover those fields first. Moreover, the probability of discovering a given area of oil field per unit effort would decline exponentially with time. Considering that, in fact, practically all the oil discovered in the forty-eight states was in giant fields and that the discoveries per unit effort had declined exponentially with time, it was apparent that the history of oil exploration could be modeled by Monte Carlo simulation of random drilling.

We programmed a computer to sample randomly the whole area of the contiguous states including all known oil fields. At the end of each unit of "drilling," (10^8 feet, or 20,000 holes about a mile deep) we determined which fields had been "discovered." In this simple way, we made ten Monte Carlo simulations of the discovery of oil as it would have occurred by random drilling at the historic rate. When the actual history of discovery was plotted in comparison, it lay within the envelope of simulated histories. In brief, the exploration for oil had been no more successful than random drilling.

The computer was searching for *area* of oil fields, which is not exactly related to the volume of oil in fields. In fact, the simulated search did much better than industry in finding some types of fields—those with large area. The largest field by area or volume in the contiguous states is the East Texas field, which was discovered by industry in 1930. In nine of ten simulations, the computer found the field before that time. (Moreover, that giant field was not actually found by the geologists of organized oil companies but by a small-time wildcatter who was drilling on a hunch.) This could have been predicted without Monte Carlo modeling. The area of the field (*f*) is 567 km^2; the area to be explored (*A*) is 4,700,000 km^2; if the number (*n*) of holes drilled is 20,000, the probability of finding the field is

$$1 - \left(1 - \frac{f}{A}\right)^n = 0.91$$

by drilling at random. In fact, industry had already drilled 300,000 exploratory holes when the East Texas field was found. The probability of *not* finding the field with that number of random tries is 2×10^{-16}. The cause for this bad luck is not wholly understood, but it seems clear that, in organized exploration by Western civilization, doing as well as pure chance may be something of an achievement.

Some leaders in the oil industry were hardly surprised, although they had not had a quantitative evaluation of efficiency before. They already knew from their own unpublished analyses that they would have done better by drilling on a grid or even at random in some unexplored provinces. It is evident that geologists and geophysicists can identify the kinds of rocks and structures in which oil may accumulate, and they are efficient at finding small oil fields in known oil provinces—which are equivalent to island clusters. Apparently, the lack of efficiency in finding giant fields derives from an institutional persistence in drilling for oil in one of the possible types of oil-bearing structures when in fact the oil in a province is in another type. Thus, having found oil in anticlines, industry might drill one anticline after another in a province where the oil is in ancient coral reefs. It is rather like generals refighting the last war. Meanwhile, the naive computer is just as apt to drill the first exploratory hole in a reef as in an anticline.

The correspondence between the model of random drilling and the actual history of oil exploration seems to justify some general conclusions regarding exploration for oil fields or islands:

1   The largest objects tend to be discovered first.
2   There is an exponential decline in the proba-
     bility of finding an object of a given size with
     a unit effort of searching.
3   Once the first object in a cluster has been dis-
     covered, the remainder are easier to find.
4   The ideas of explorers can greatly affect their
     chances of success.

EUROPEAN DISCOVERIES

As far as explorers are concerned, islands differ in one fundamental way
from oil fields—they are capable of killing the unwary. Thus, the attitudes
of sailors regarding uncharted waters are always mixed. In the late nineteenth century, navigational charts were full of chimerical islands because
every possible hazard to navigation, however questionable the information
suggesting its existence, went on the charts. As late as the 1960s, charts of
the South Pacific were full of the notations "P.D." for "position doubtful"
and "E.D." for "existence doubtful," regarding rocks and shoals. The only
prudent course for a captain was to avoid the site of any possible hazard. So
it has always been, except for those few surveying and oceanographic ships
whose job it is to deliberately seek and survey such hazards or disprove
their existence. It took Western civilization about 1500 years to discover
all the oceanic islands, and it appears that Captain Cook and his lieutenants were almost the only people in all that time who took their surveying
job very seriously.

The probability that an island will be found by sailors depends on its
size, its distance from a home port, the number of voyages from the port,
the freedom of action and spirit of adventure of captains, the likelihood of
ships' being driven long distances by storms, and so on. All in all, it is not
surprising that the largest oceanic volcano, Iceland, was the first to be
discovered, in the fourth century A.D., by the Norsemen, who lived not
far to the east. They colonized the island by the ninth century and roamed
the northern seas—which contain few oceanic islands.

The next phase of discovery was in the fourteenth and fifteenth centuries, when Portuguese, Spanish, and other European explorers began to
seek a sea route to the spice and silk of the East. Just as Columbus accidentally found the vast area of the Americas, so others sighted tiny oceanic
islands or ran aground on them. In 1420 the Portuguese Zarco discovered
the Madeira islands, for the last time, when storms drove him west from
his exploration of the coast of Africa. A Genoese map of 1351 shows that
contact had been made before—the islands are only 670 km west of Africa

Global wind patterns that determine sailing courses and migration paths.

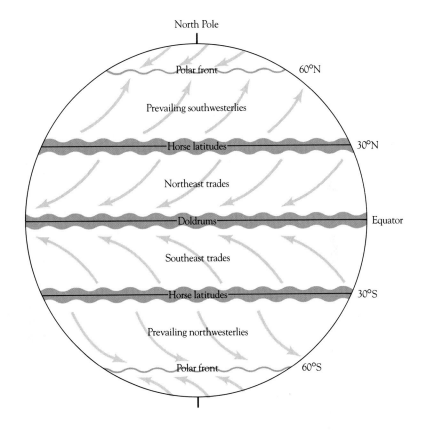

North Pole

Polar front   60°N

Prevailing southwesterlies

Horse latitudes   30°N

Northeast trades

Doldrums   Equator

Southeast trades

Horse latitudes   30°S

Prevailing northwesterlies

Polar front   60°S

and the Straits of Gibraltar. The Azores, even farther west, were already known to the Carthaginians, who left coins, and Arabian geographers. They were discovered for the last time in 1432, when Van der Berg was driven on the islands by a storm. Although the Azores are in three widely separated groups, all nine islands were found and some even colonized by the Portuguese within twenty-five years. We may generalize that, like oil fields, once one of a cluster of high islands is found, the rest will be discovered quickly if there is any desire to do so. Among other reasons, each high island is commonly visible from the peaks of another in the cluster.

As the Europeans sailed farther south, further discoveries were made apparently for the first as well as the last time by man. These included the cluster of the Cape Verdes in 1456; the tiny, isolated, midocean islands of Ascension, in 1501, and St. Helena, in 1502. Clearly, the explorers were tacking far into the Atlantic to follow the latitudinally zoned winds. The Portuguese reached oceanic islands in the Indian Ocean soon after: Mauri-

The cliffs of ironbound St. Helena.

tius in 1505, and Reunion in 1513. All of the islands discovered to this time had several features in common. They were high volcanoes, active or dead, uninhabited, and wholly lacking gold, diamonds, or anything else offering quick profit. Some were ironbound by great cliffs but even these had a few protected anchorages and fresh water, so the islands had some use. Moreover, being high, they were visible from great distances and thus hardly hazardous to navigation.

So when Magellan entered the Pacific, in 1520, he had some knowledge of oceanic islands. We may pause to consider what else he knew and his situation. He knew about the trade winds. After beating his way through the straits that bear his name it could hardly have escaped his attention that he was in the wrong latitude to sail west. Not to mention that the known riches of the East were in the Northern Hemisphere. His ship was marginal for the voyage and his supplies were already low. Considering all these factors, his only logical course was to sail northwestward until he reached the tropics and the gentle, persistent easterlies of the trade winds. This he did.

The state of the science of navigation in Magellan's time enabled him to determine latitude at sea, but not longitude. Indeed, in those days before surveying by triangulation, no one knew longitude very well on land, either. The course being steered and speed made good through the water could be measured, but wind and sea drift were always uncertain, and often hopelessly so after a series of storms. As a consequence, the longitudinal positions of ships not infrequently were in error by hundreds of kilometers and occasionally by more than two thousand kilometers. Not until Captain Cook's time, in the late eighteenth century, were nautical chronometers accurate enough to permit determinations of longitude. Even two centuries after Cook, positioning errors of 15 km to 30 km were common in celestial navigation. Not until the invention of electronic and artificial satellite navigation in the 1960s and 1970s did a ship at last know where it was most of the time. Then, naturally, almost everything that had been discovered had to be relocated.

Explorers of the Pacific

Magellan made the first European discovery of a Pacific island on 24 January, 1521, but we do not know which one, for lack of a longitude. From its latitude and its description as a low island fringed with trees, we know it was an atoll in the northern Tuamotus, but whether Fangahina, Angatau, or Pukapuka is uncertain. (This ingenious method of identifying island discoveries by combining latitude and island description was developed by Andrew Sharp, and his chronology is used here.) Magellan saw no sign of inhabitants and could not anchor on the steep coral bottom, so he sailed on. He had found a small, low, surf-bound, valueless hazard to navigation. The last oceanic island in the main Pacific basin, the 267th, was discovered in 1859 by Captain N. C. Brooks of the Hawaiian barque *Gambia*, and for some time the uninhabited island took the name of the ship. Then it was renamed and in due course gave its new name to the most famous naval engagement of World War II, the Battle of Midway.

The whole period of discovery lasted 338 years. If we divide it into 50-year intervals, it is evident that there were two major phases of discovery. The first began with 32 discoveries before 1550 and tapered off to the interval 1651–1700, when only three islands were discovered. The second and greater phase began with 12 discoveries in 1701–1750 and peaked at 113 in 1751–1800. Two-thirds of all the islands were discovered in the century beginning with 1751. The variations in discovery rate were due to improvements in ships and navigation, concern with hazards, variations in the frequency of voyages, and changes in the motivation for voyaging. Of these the last apparently was dominant.

It appears that voyagers from 1521 to 1700 viewed Pacific atolls and volcanoes with more fear than hope. The famous explorers Quiros, Mendaña, Schouten, and Le Maire all followed routes of easy sailing, west on the southeast trades and home either around the world or east on the westerly winds in high northern latitudes. Sir Francis Drake sailed across the Pacific from California, presumably on the northeast trades, and reported no islands at all. Thus he confirmed the wisdom of the Spanish

Rebotel Reef, in the Palau Islands, is the kind of navigational hazard that Pacific explorers avoided for centuries.

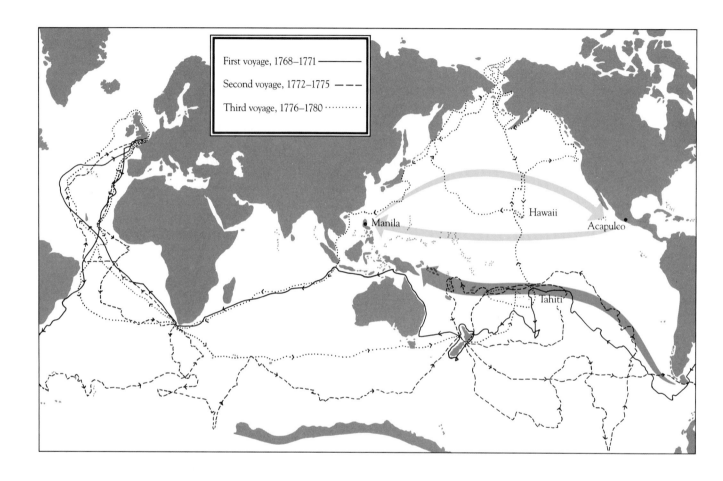

The voyages of that uniquely determined explorer Captain James Cook. Most so-called explorers followed the safe and easy highway in the South Pacific. The prudent Spanish merchants followed the safe loop in the North Pacific.

conquerors who, beginning in the late sixteenth century, sent galleons from Acapulco to Manila along 13°N latitude and back at 40°–60°N latitude. For centuries they sailed the same route because exploration had shown it to be safe. It was all as routine as the P&O sailings from England to India in the days of empire, although the best accommodations were not POSH but SOPH. Naturally, the Spaniards discovered few islands as they sailed in a vast loop around the unknown Hawaiian Islands.

With the dawn of the eighteenth century came a thirst for geographic knowledge, science, and, perhaps more important, a final hope for territorial expansion on a continent thought to lie in the South Pacific. The British troops who surrendered at Yorktown later in the century played a tune, "The World Turned Upside Down." That is what theoretical geographers once thought would happen if the many continents in the Northern

Hemisphere were not balanced by equal continents to the south. Thus a new wave of explorers moved through the Pacific basin. Roggeveen with two ships found eleven islands from isolated Easter through the Tuamotus, Society and Samoan archipelagoes. Byron, Wallace, Carteret, and Bougainville followed with comparable discoveries. Unfortunately for territorial hopes, they more or less followed the same old explorers' turnpike.

Enter the incomparable Captain James Cook, who made three voyages from 1768 to 1779, when he was killed in Hawaii. Even he followed the turnpike on his first voyage, but thereafter he followed logic and took the west winds to crisscross the South Pacific. In this he was preceded by Tasman, who sailed on the westerlies south of Australia in 1642 and (after the Maoris and the kiwis) discovered New Zealand. Cook came the same way, and between 1772 and 1775 he eliminated the possibility of a southern continent outside polar waters. He did a similar search of the North Pacific on his last voyage; he bisected the Acapulco–Manila loop and found the Hawaiian Islands.

In the central Pacific basin, Cook found and surveyed 30 islands. Through his unique influence and training, his lieutenants and their lieutenants, seemingly everyone associated with him, continued to explore. His lieutenant Clerke found the last two high Hawaiian Islands. A decade later, his former navigator, Captain Bligh, discovered two islands with HMS *Bounty*. When the mutiny occurred, Bligh and the loyal sailors were placed in an open boat. They then made the longest recorded voyage in such a boat, all the way to Batavia, seldom touching land for fear of the Melanesian cannibals, who even paddled out from shore to intercept them. In the midst of all these hardships and perils, Bligh discovered—and surveyed one side of—eleven islands in the Fiji and Banks groups. (Cook had once remarked that to survey an island he frequently had to expose his ship on a lee shore, which was contrary to all his training. He did so because the Admiralty had sent him out not to preserve his ship but to survey.) His chief mutineer, Lieutenant Fletcher Christian, discovered fertile Raratonga (and the Raratongans) with *Bounty* before reversing course and eventually burning the ship off the landing on isolated, uninhabited Pitcairn. To complete this log, Captains Edwards and Oliver, searching for the mutineers, discovered three more islands in the central Pacific and four more among the continental islands of the Solomons.

The Efficiency of European Exploration

The oceanic islands of the main Pacific Basin east of the island arcs comprise 184 atolls or rocks barely above sea level and 83 high islands, including elevated atolls. The distinction is made between high islands and low because height is what determines how far an island can be seen—its

The western shore of Pitcairn Island has the high cliffs typical of even very young volcanic islands in reefless seas.

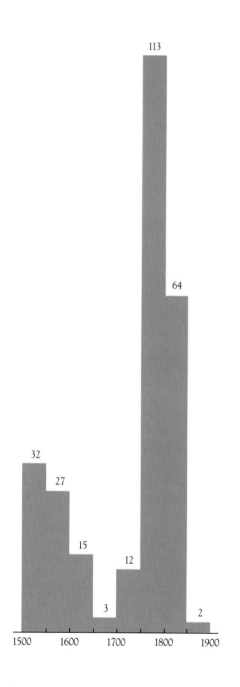

The number of islands discovered in the Pacific in each fifty-year period from 1500 to 1900. Clearly by 1700 no one was looking for islands.

"size," for the purpose of discovery. Thus, the history of discovery of the finite population of high and low islands in this circumscribed area may be compared to the better known history of the population of giant and small oil fields in the United States.

In the first phase of exploration, from 1500 to 1700, the number of islands discovered per 50 years systematically declined. Presumably this reflects both a decline in interest in the Pacific and the fact that its Spanish masters were content to conduct commerce along known, safe routes. As far as island discoveries go, therefore, the heroic first phase did not amount to much. Nothing of value was thought to exist in the main Pacific basin, so it was pointless to search for islands.

In the second phase, beginning in 1700, discoveries per fifty years averaged about four or five times the rate in the previous two centuries. However, within this phase, particularly from 1760 to 1860, there was hardly any systematic trend in the rate of discovery per decade until the 1830s, by which time almost all the islands had been discovered. Even if the discoveries by the unique Cook are eliminated, the rate varied randomly from 10 to 29 per decade for 70 years despite the almost complete exhaustion of the finite population of islands available for discovery. Random searching at a constant rate would have produced an exponential decline in the rate of discovery. If, indeed, random searching is an appropriate model for European exploration for Pacific islands, there must have been a balancing exponential increase either in the rate of exploration or in its efficiency.

It is easy enough to devise simple models of random searching and apply a Monte Carlo method to generate simulated histories of discovery. All that is necessary is to determine the size and position of the targets and then have a computer run straight lines or random sailing courses through the search area. Two models might be necessary because the size of the target varies with the objectives of the searcher. We define "finding" as seeing an island. A voyager who is trying to avoid islands discovers one only if it chances to come in sight. However, islands give many signs of their existence besides being visible. The orographic rain clouds that tower above high islands are often visible long before the island comes in sight. Likewise the milky blue-green color of a still-invisible atoll may be reflected on the clouds of the trade winds. Inasmuch as the temperature of a lagoon is higher than that of the surrounding water, the pattern of the little tropical clouds over the lagoon may also be revealing. Land birds, floating vegetation, seals, wave and swell patterns, even smell can indicate the nearby presence of land yet undiscovered. Thus, one computer program would sail straight on and the other would begin a box search until the discovery was made. The latter would be much more successful because, effectively, it would be seeking much bigger targets.

Orographic rain clouds over Palau. The high, stationary clouds presented large targets for explorers seeking islands.

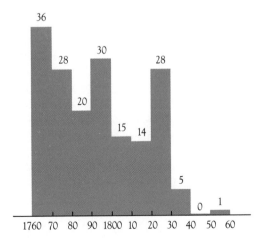

Number of islands discovered in the Pacific per decade, 1760–1860. Evidently, there was no systematic trend.

The problem in measuring the rate and efficiency of European exploration is that the total length of all Pacific voyages is unknown. Thus, there is nothing comparable to the total length of exploratory drilling for oil fields. Consequently, the actual efficiency of exploration—number of islands discovered per unit effort—cannot be determined for comparison with random searches.

What can be determined from Sharp's chronology of discovery is how many islands were found on each voyage that found any islands at all. Consequently, it is possible to see how this number varies per unit effort, even though the sample is very small compared with all the voyages that discovered nothing at all. An appropriate measure of success would then be the excess number of islands (that is, in excess of one) discovered per successful voyage per ship. Small though it is, the sample suggests that this number declined exponentially from 2.5 in the first fifty-year period to 0.5 in the last fifty-year period of the first phase of exploration. Considering that two-thirds of the islands were still undiscovered in 1700, it appears that either no one was looking for them or that the searches had bad luck comparable to that in the search for the giant East Texas oil field.

The more voluminous data for the period from 1760 to 1840 are quite consistent in suggesting that chance was a major factor in the discovery of

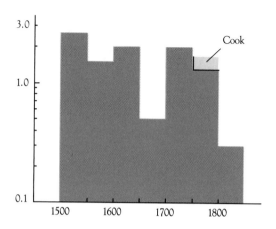

The number of excess islands discovered per successful voyage per ship was relatively constant in the fifty-year periods from 1500 to 1800.

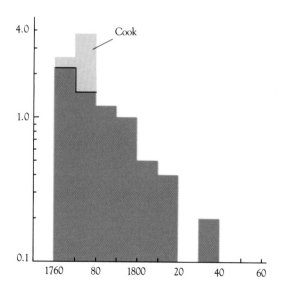

The number of excess islands discovered per successful voyage per ship per decade during the intense second phase of exploration. The number declines exponentially as it would in a random search for a finite population.

Pacific islands. Excluding Cook's discoveries, the excess number of islands discovered per successful voyage per ship declined exponentially from 2.2 to 0.2 in each ten-year period, with only a gap from 1820 to 1830 to mar the picture. Even including Cook, whose searches were far from random, an exponential decline is apparent. It appears that in the second phase of Pacific exploration the actual rate of discovery per unit effort declined as it would have in a random search. Thus, if the rate per decade remained fairly constant, it must have been because the amount of searching, whether deliberate or random, increased exponentially.

In each phase of exploration, the high islands were found generally before the low ones. This is best seen in the last century of discovery. All but two of the high islands were found by 1800 and the last, Rimatara, by 1811. In contrast, more low islands were found in the 1820s than in any other decade in the two phases of exploration. Atolls continued to be found for 48 years after the last high island. It seems that, like the discoverers of oil fields, European and later American explorers found the big targets first.

The first high island to be discovered in the Pacific region of interest here was Ponape, 786 m high, in 1529. Ponape is one of three widely separated high islands among the abundant atolls and drowned atolls of the Caroline group. The atolls surrounding Ponape were discovered in 1529, 1568, 1773, and 1824. It is evident that atolls can easily escape notice. There are curious anomalies in the other direction. We may recall that Darwin on HMS *Beagle* missed seeing the nearby phosphate island of Makatea but saw Tahiti in the distance at dawn; yet Makatea, only 110 m high, was discovered in 1722 and Tahiti, 2228 m high, not until 1767. In general, the high islands of the eastern Pacific were discovered before the far more abundant and clustered high islands to the west. The Galapagos were the first group found; all twelve were discovered in 1535. The eastern islands along the return loop from Manila were discovered early, even the tiny but high spire of Alijos Rocks off Baja California was found in 1558. The only other group found in the sixteenth century were the southern Marquesas. Only eastern high islands were discovered in the seventeenth century, and the last two of these sparse eastern islands, Easter and Sala y Gomez, were found by 1722.

The Society and Samoan groups, if not all of their islands, were discovered before Captain Cook's time, but thereafter most of the high islands were discovered by him and his lieutenants. After their time, little was left. The remote phosphate islands of Nauru and Ocean were almost the last, in 1798 and 1804 respectively. Curiously, on those islands was what the Spaniards and their successors despaired of finding—a fortune in ore.

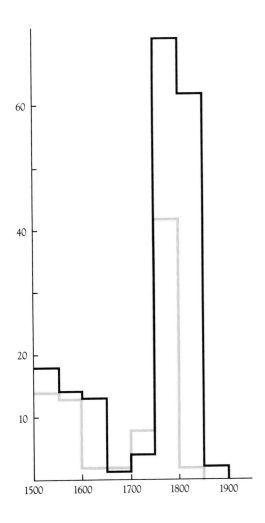

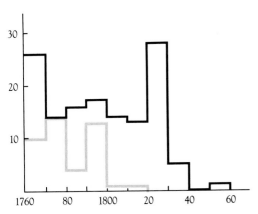

The rates of discovery of high (color) and low (black) Pacific islands by fifty-year period (above) and by decade in the period 1760–1860 (below).

POLYNESIAN COLONIZATION

The Europeans who entered the Pacific for more than 300 years came from an unwashed, polluted, disease-ridden culture that had passed from the Bronze Age to the Iron Age several millennia earlier, a culture that had cannon, cathedrals, and the printing press. The people they found on the tiny isolated islands were clean, healthy, and generally friendly, and they seemed exceedingly handsome to sailors long at sea. Polynesians had a materially simple stone-age culture with only three domestic animals—the pig, cat, and the chicken. There were no wild mammals or reptiles to hunt or defend against. There were no machines, no wheels, no pottery—because atolls and volcanic islands lack clay. Food was abundant but the variety of fruits and vegetables limited. The sea provided a wide variety of limitless protein. These cultures had complex social structures with kings and nobles, property rights, warfare, and religion. Islanders built temples in the form of uncemented but sand-filled stone platforms. On some of the high islands, they ornamented the temples with large stone statues resembling the well-known ones on Easter Island. The statues were carved of volcanic tuff, easily worked with obsidian or pitchstone hand axes, so little technology was involved.

The first scientist to encounter Polynesian culture was Joseph Banks, who was with Cook on his first voyage. The future Sir Joseph would long be the President of the Royal Society, but on Tahiti he was a young man in paradise. He studied the botany, but his journals make it clear that he spent more time enjoying than analyzing the complaisant society of Tahiti. Scientists who later visited the islands began to devote themselves to the origin, history, and culture of the Polynesians. Evidence was derived from oral traditions, physical anthropology, serology, domestic plants and animals, artifacts, and analysis of cultural evolution. A very strong consensus among diverse specialists was that the Polynesians had come from southeast Asia, probably from what is now Indonesia. If so, they had peopled the Pacific by sailing into the trade winds. The magnitude of their achievement can be perceived from the number and desperation of the European attempts to deny that it happened.

A long European tradition proclaimed that in general one did not sail east into the trade winds. The European way to go east was to do so on the westerlies at high latitudes, but to do that was a comparatively difficult technical feat. Europeans and, later, Americans could not believe that a Stone-Age culture was capable of a large-scale migration by either route. Almost every conceivable alternative was proposed, and it seemed that the wilder the idea the greater its popularity. It was proposed, by a scientist, that the sailing simply had not taken place. A gigantic, Pacific-wide

The ecologically well-adjusted Polynesian culture as portrayed by John Webber, who was with Captain Cook on the third voyage.

Hawaiian warrior, engraved from a sketch by John Webber.

continent had been submerged and what were now islands had once been its mountains. The Polynesians had walked on their migration and merely retreated to the peaks when submergence separated them. A drowned continent was a popular idea among biologists, particularly botanists, but few would have agreed that it had submerged while man was on earth. A nonscientific enthusiast gave the continent a name, "Mu," and wrote several books comparing its history with that of the other imaginary sunken continent—Atlantis.

Some people allowed migration by ship, but surely not by a Stone-Age culture. It followed that the Polynesians were little more than the feral remnant of a high culture. But what high culture? Among the possibilities considered were a lost tribe of Jews, or of Aryans, or the mysterious but doubtless technologically mighty inhabitants of Mu. In yet another interpretation, the Polynesians themselves did migrate by sea, but this was not much of a technical achievement because it was done downwind on the trade winds from South America. Thor Heyerdahl demonstrated that the voyage could be made on a properly provisioned raft that was towed across the near-shore currents. His account, *Kon-Tiki*, went through seven printings in its first year, 1950, and ultimately more than twenty-five printings.

The idea of a *simulated* Polynesian voyage from South America would not have surprised Sir Peter Buck, who in 1938 published the concept that there were real voyages. Buck, however, assumed that the voyagers had first sailed from Polynesia to South America. The return downwind would

Polynesian double-hulled canoe from the Society Islands.

then have been easy. Sir Peter Buck was a Maori, born Te Rangi Hiroa, who left a position as a Maori medical officer to pursue the origins of his people. Speaking a Polynesian dialect as his mother tongue, he made extensive use of interviews to obtain oral traditions, histories, and genealogies, some of which went back 92 generations. With Buck the pendulum at last swung. The title of his book *Vikings of the Sunrise* referred not to the antecedents of the Polynesians but their abilities as sailors and navigators. It went through two editions and additional printings and has become widely accepted, especially in Polynesia. He visualized fleets of double-hulled sailing canoes that set sail, according to plan, bearing hopeful emigrants and the provisions to support them. On the broad platforms between the twin hulls were the domestic animals, plants, and seeds to establish new settlements. The voyages counted on rain to supplement water, and upon fish to supplement food.

The great canoes were seen and illustrated by early European voyagers, so Buck's interpretation of Polynesian history began on firm ground. He knew the South Pacific well and was scornful of the "nonsense" in print about the impossibility of sailing east in the latitude of the trade winds. The trades sometimes ceased and were replaced by westerly winds from time to time. He cited the experience of the pioneering Christian missionary John Williams, who sailed east from Samoa to the Cook Islands on a straight course without changing tack. In any event, sensible sailors preferred to explore by beating against prevailing winds because, if no new island was discovered, they could speed home to food and water. The only weak link in this appealing history of noble human achievement was the possibility that the island hopping was accidental. Perhaps the Polynesians populated new islands only when their sturdy canoes were driven who knows where by great storms. Buck cinched his analysis by pointing out that, although women swam, dove, fished and sailed, it was only within lagoons. They did not accompany men in fishing in the open sea where they could have been blown away. No women, no new colonies; it was as simple as that. If the women were at sea, it could only have been with the great colonizing fleets.

Sir Peter Buck had painted an attractive picture, consistent with mainstream science and based on a personal compilation of oral history in the 1930s. In 1956, Andrew Sharp pointed out that the picture was not consistent with earlier observations of Polynesian culture. Sharp observed that once the Europeans arrived they grossly changed Polynesian life. Polynesians on some islands were almost exterminated by European diseases. Cultures were rapidly corrupted, as they were all over the world, by the awesome European technology. The isolated Polynesian society was exposed to the world. For example, Captain Cook's Tahitian translator,

The Polynesian culture was quickly intermixed with and overwhelmed by European culture. Captain Cook's Tahitian interpreter, Omai, was painted in London by Joshua Reynolds. Within a few years, Omai and other Polynesians had returned home with new versions of Pacific geography and history.

Omai, had spent two years in London before sailing on Cook's third voyage. Even within the Pacific, Polynesians traveled with the Europeans and, moreover, could learn of many islands with which they had not necessarily been familiar. Thus the memories and, possibly, the traditions of Polynesians after the great discoveries of 1760–1780 were suspect.

It is prudent, therefore, to go back to European journals and logs of voyages to Polynesia before any significant changes occurred. The first scientific voyage was Cook's on *Endeavour* in 1768–1771. Cook, Banks, and Solander were all curious and qualified observers, and the journals of the first two have something to say about Polynesian origins. Banks believed that the Polynesians had come from the west because of their language and their domestic plants and animals. Cook, the master mariner, saw not the slightest problem in accepting that the migration was against the trade winds. He found that the inhabitants of the Society Islands were familiar with islands "laying some 2 or 300 Leagues to the westward of them." He assumed, in those days early in the second phase of European exploration, that island succeeded island to the west. Thus the inhabitants of the islands west of Tahiti would in turn know of the islands west of them, and so "we may trace them from Island to Island quite to the East Indies."

By his third voyage (1776–1779), Cook had more data and a more complete hypothesis of Polynesian migration. Polynesia was divided into two main regions: western Polynesia, consisting of the Tonga, Samoa, and Fiji groups, and eastern Polynesia, which included the Society and Tuamotu islands. The Polynesians told the early explorers that *deliberate* voyages were made only within the two regions. How voyages were made between groups or to islands outside the groups was suggested by what Cook learned at Atiu, in what are now the Cook Islands. Omai, the interpreter, found three of his fellow Tahitians on Atiu, 1100 km from home. They were the survivors of a party of twenty who had expected to have a brief sail from Tahiti to Raiatea, barely over the horizon at sea level. Cook knew of many other accounts of accidental voyages such as one in 1696, when a large canoe was driven by storms from the Caroline Islands to the Philippines, 1800 km away. Men, women, children, and babies survived. Cook reasoned that such accidental long voyages by family and tribal groups attempting easy interisland trips

will serve to explain, better than a thousand conjectures of speculative reasoners, . . . how the South Seas, may have been peopled; especially those [islands] that lie remote from any inhabited continent, or from each other.

In short, Cook proposed that the islands were peopled not by hypothetical great fleets of migrators but by an essentially random search, which was still going on.

Andrew Sharp fleshed out this skeleton of an idea with data from the time after Cook's death, on his third voyage. Accidental voyages were more frequent toward the west because of the normal trade winds. However, there were many also to the east during lulls in the trades or, more commonly, when gales or typhoons overwhelmed the normal weather. For example, a canoe-load of people from Manihiki in the Northern Cooks survived an accidental voyage of 1100 km to the southeast to Aitutaki in the Southern Cooks. Another important influence on the probability of long accidental voyages is the frequency of inter-island travel by groups of men and women. Sharp showed that family and group voyages to nearby islands were commonplace in the nineteenth century just as they are now. The population of one pair of islands moved *en masse* back and forth between them every few years; their use of the islands was rather like crop rotation. Other people would go off to visit family connections on nearby islands; or to colonize a less desirable and thus unoccupied area of a nearby island.

A question might be raised about the probability that a group of families would survive for weeks when they had supplies on board for only a day or two. The probability cannot be assessed; perhaps most of those swept away were drowned or died of exposure, hunger, and thirst. Nonetheless, successful storm-driven, accidental voyages may have been numerous enough to populate the islands. In any event, the ability of the ancient Polynesians to survive at sea defies the modern urban imagination.

Some faint idea of what can be done is provided by the little book *Survival on Land and Sea*, prepared for the U.S. Navy by the Ethnogeographic Board of the Smithsonian Institution. I have read my copy many times since I received it on shipboard in 1944. After a few special sections about not drowning in a parachute and about surviving under burning oil from a ship, it presents a manual for staying alive in a life raft that would apply to anyone adrift. You can live for weeks without food and 8 to 12 days without water. A pint of water a day keeps you fit if you are not active. Moreover, fish hooks can be made from many materials, including wood, and fish line from cloth or rope. Small pelagic sharks collect under and around boats, and birds, flying fish, and squid may land aboard. Rain can be expected to provide water; and potable water, rather like oyster juice, can be squeezed or chewed from freshly caught fish. Exposure can be a problem; I would never abandon ship without a hat. However, awnings can be improvised and clothes minimized, so that perspiration is free to evaporate but the sun is still screened. Clothes should be dipped in

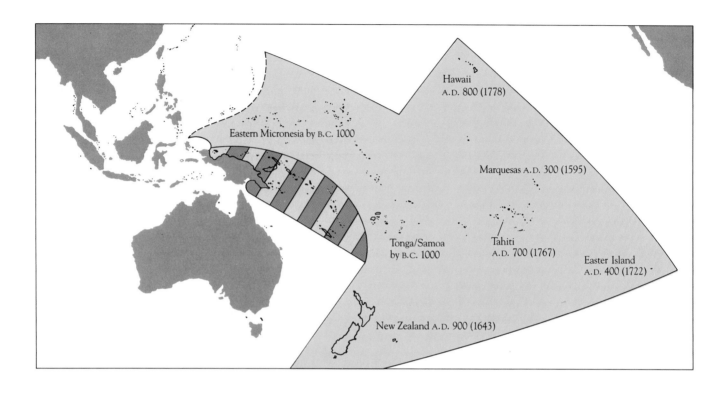

Hawaii
A.D. 800 (1778)

Eastern Micronesia by B.C. 1000

Marquesas A.D. 300 (1595)

Tonga/Samoa
by B.C. 1000

Tahiti
A.D. 700 (1767)

Easter Island
A.D. 400 (1722)

New Zealand A.D. 900 (1643)

Dates of discovery of central Pacific islands by Polynesians (Europeans). The area of Polynesian settlement after 1000 A.D. is shown in color. North and east of New Guinea, Polynesian settlements coexisted with Melanesian cultures.

salt water to provide cooling by evaporation, although care should be taken not to be chilled.

With this kind of information, and determination, young men from the fields of Iowa and the streets of Chicago have survived for weeks in open boats and rafts. The Polynesians were as nautical a culture as ever existed, swimming like otters, sailing from infancy, and fishing for a living. They started on their inter-island cruises or inter-archipelago expeditions with just the sorts of gear, rigging, and sails that were most useful for long survival at sea. Even after destructive storms, enough voyagers could have survived to people the Pacific.

Thus, the random-voyage hypothesis seems entirely adequate to explain the peopling of the Pacific, although it has evoked a mixed response. A troubling aspect is that it seems to diminish the Polynesian achievement—in fact, it is not known that large-scale planned migrations did not occur. However it was accomplished, what was the chronology of Polynesian exploration and colonization?

The Polynesian culture apparently developed among people who migrated from Indonesia through Melanesia to the Samoa-Tonga region per-

Stone heads on Easter Island mark the easternmost occupation of Polynesian colonists. However, plants and mineral specimens indicate noncolonizing voyages on to South America.

haps 3000 years ago. Yet radiocarbon dating so far has not shown any occupation of the central South Pacific islands until the first millennium A.D. Presumably, human waifs were coming and going east with storms and home again with the trade winds. Given a seagoing people in western Polynesia for at least 1000 years, it seems impossible that they did not accidently learn of the islands farther east. Nonetheless they neither deliberately nor accidentally populated Tahiti or the other islands. Could it be that home in Tonga or Samoa was so ideal that every group of shipwrecked waifs merely built a new boat and sailed back on the trade winds?

Apparently, something changed, perhaps population pressure, and Polynesians occupied the incredibly distant Marquesas islands about 300 A.D. They were on Easter Island by 400 A.D. and throughout the Society, Tuamotu, Austral, and eastern Cook islands by perhaps 700 A.D. In 800 A.D. they probably were in Hawaii and a century later in New Zealand. The Polynesians also discovered, although they did not permanently occupy, numerous isolated islands—probably more than once. There are ancient ruins in the interior of some high islands in the Carolines, and an abandoned temple on Pitcairn. Skeletons, artifacts, and ruins are spread from the Line Islands to Henderson, which is southeast of Pitcairn. Polynesians not only reached the Galapagos but somehow made contact with South America, whence the sweet potato was brought to New Zealand.

In sum, Polynesians discovered and colonized the islands of the open South Pacific as well as Hawaii in about 600 years. They had gone everywhere from New Zealand to South America. It took Europeans with much better ships almost half as long just to find the islands, and they never have colonized many of them. We have no way of knowing how much of the colonization was deliberate and how much accidental, but regardless of how the Polynesians peopled the Pacific, it seems to have been reasonably efficient—pigs, chickens, and all. Considering that in exploration it is no small achievement to do as well as pure chance, there is no way to diminish the greatest maritime feat in human history.

POPULATION BY PLANTS AND ANIMALS

A large number of biologists have studied island life in the past hundred years, including many specialists in subjects that rarely overlap. Inevitably there is a great diversity of apparently conflicting evidence and thus a range of opinion on how plants and animals populated oceanic islands. Even so, there is agreement on the one point that is most controversial regarding human exploration: Plants and (non human) animals found the islands accidently, without intent, and entirely according to the laws of chance. One might reason that chance would favor those waifs biologically more capable of dispersal, like the Europeans and Polynesians, who were culturally prepared to discover oil fields and islands. We have seen that such human discoveries may be inevitable even if there is a large element of chance. Regarding other species, biologists also seem to have achieved a consensus that organisms capable of long-distance dispersal are more apt to be on an island than not.

On many other points, controversy continues. For example, faunal affinities indicate that different organisms reached such islands as Hawaii from different continents, but when and by what routes is less certain. In the last chapter of this book, we shall view island life in the light of plate tectonics and insular geology, which have some bearing on the history of dispersal to islands. Here, however, we shall focus on the paths by which colonizing plants and animals reached the islands. Many, perhaps all, possibilities have enjoyed scientific support; these include migration across former continents or former linear continental fragments called "land bridges," hopping along former island chains, and simple dispersal to the islands as they are now distributed.

The hypothesis that ocean basins and continents were not permanent had widespread support from the early nineteenth century until fairly recently. Many of the most eminent geologists believed that dry land had been where the ocean basins are now and that subsidence had merely transformed one into the other. Thus, biologists could cite expert geological opinion to explain the modern distribution of plants and animals. The geological evidence that was explained by the hypothesis was of two types,

Granite outcroppings amid the coral sands of the Seychelles Islands in the Indian Ocean. The Seychelles are a tiny fragment of drifting continent quite unlike the volcanic and coral islands typical of ocean basins.

and by the late nineteenth century the facts were hardly in dispute. First, marine fossils and sedimentary rocks occur widely on continents, including what are now the peaks of the highest mountains. Clearly, the land that can be seen has once been the sea floor. It once seemed only reasonable that the sea floor, which could not be studied in such detail, might once have been land. Second, Paleozoic and early Mesozoic fossil assemblages of the Atlantic coasts of Africa and South America are very similar, and this is true of Northern Europe and North America as well. Furthermore, the sequences of sedimentary rocks on opposite sides of the Atlantic are also very similar, and the geological structures of the two coasts trend out to sea. It is obvious that at one time Africa and South America, for example, were connected by dry land.

To plant geographers, the idea of foundered continents was particularly attractive. J. D. Hooker, Darwin's friend and one of the earliest supporters of evolution, did not see how the "peculiar endemics" of insular floras could be explained by random dispersal over water. Moreover the insular floras reflected a "far more ancient vegetation than now prevails on the mother continents." All manner of problems about dispersal from continents to isolated islands were identified by botanists and other biologists as well. These problems posed no difficulties if the islands were merely peaks of foundered continents. All was explained by land distributions that had now vanished. On the other hand, all these problems were acute for a second group of biologists who believed that ocean basins and continents were not interchangeable. To be convincing, they would have to prove that long-range dispersal over water was not only possible but going on now.

The pioneer in the second group of biologists was Charles Darwin, whose reasoning derived from his hypothesis on the origin of atolls. He had proposed that the coral atolls were reefs built on the tops of isolated submarine volcanic edifices. If the bases of the volcanoes had once been connected by dry land, as parts of a continent, the coral would have grown up like the Great Barrier Reef off Australia, only the Pacific reefs would have been even more extensive. Furthermore, with very few exceptions, the only rocks found on islands in deep ocean basins are volcanics, such as basalt, and coral limestone. If the islands were peaks of foundered continents, they should be like the peaks of unfoundered continents. Many should have outcrops of Paleozoic or Mesozoic fossiliferous sedimentary rocks like the Alps or Himalayas, or granite and metamorphic rocks like the Sierra Nevada. Darwin said that they did not, and if everyone had accepted his conclusion, the foundered-continent hypothesis might have been abandoned. However, as Darwin himself noted, the Seychelles Islands, rising from the deep Indian Ocean, are in fact coarse granite. Moreover, many of the pioneering geologists who followed the discoverers of islands seem to have had very bad luck in sampling and describing rocks. On many islands they found what were interpreted as metamorphic and igneous rocks more like continental granite than oceanic basalt. How they did this on what are now obviously youthful volcanoes rising from oceanic crust is mystifying to nonpetrologist. However, the samples were few and the interpretations made in good faith, so as late as 1950, Darwin's conclusion was based on evidence that was widely perceived as equivocal.

After Darwin's ideas were published, the *Challenger* expedition found that the deep sea floor is covered with red clay and globigerina ooze. A. R. Wallace pointed out in his book *Island Life,* published in 1880, that rocks made of such materials do not exist on continents, and this suggested that continents and ocean basins are permanent. When Wallace had sent his first, brief manuscript outlining the theory of evolution by natural selection to Darwin, he had believed the foundered-continent hypothesis. This was hardly surprising, because much of his field work was in the Indonesian islands, which are in fact continental in composition and arise from a shallow continental shelf. In times of lowered sea level, animals could migrate about with dry paws. Darwin wrote to Wallace that he agreed with everything that Wallace proposed except for the populating of islands in the deep sea. On that point Darwin would defend his own views "to the death." Wallace soon appreciated the difference between continental and oceanic islands and supported Darwin, but other scientists did neither.

Further evidence for the permanence of continents came not from the tiny oceanic islands but, like the *Challenger* data, from the broad, deep sea. Geophysicists would show that continents and oceanic crust are too different for one to be changed into the other. By about 1900, O. Hecker had made enough measurements to show that the Atlantic, Indian, and Pacific ocean basins are as close to isostatic equilibrium as the continents are— both types of crust float buoyantly on denser material below. Thus, the ocean basins, which ride much lower than the continents, must be made of much denser rock. In the 1950s, Russell Raitt and Maurice Ewing, among others, began to measure cross sections of the oceanic crust from ships by explosion seismology. They discovered that the standard oceanic crust is much thinner than the standard continental crust and that crust of intermediate thickness is very rare. The result of half a century of geophysics at sea was a complete confirmation of Darwin's conclusion that ocean basins are not foundered continents. What then of the compelling evidence that Africa and South America had once had a land connection? Alfred Wegener had explained it all in 1915 by continental drift. The stratigraphic and paleontological evidence of trans-Atlantic linkages was undisputed, but it now had no bearing on the dispersion of animals and plants to oceanic islands.

If Darwin did not immediately convince everyone about the populating of islands, it was not for lack of his usual valiant try. He conducted a lengthy series of experiments to determine how long seeds and plants would float in sea water and still be fertile. Ripe hazel nuts, he found, sank immediately but if dried first they would float 90 days and still germinate. Dried asparagus with berries floated 85 days, and so on. He also amassed an enormous collection of observations of plant and animal dispersal. Coconuts drift across oceans, and West Indian beans regularly beach on Scotland. Birds cross oceans and carry fertile seeds in their crops. A blob of mud from a partridge's leg contained the seeds of 82 plants of five species. These experiments and observations proved that a surprising range of plants and animals could survive long-distance transportation by air or sea and reproduce on islands.

Darwin did not show that breeding pairs or genetically diverse groups of mammals or reptiles could populate islands. But there are no mammals and few reptiles on oceanic islands, except for those that were brought by people. Indeed, a correct explanation of the origin of insular populations must include a filter that eliminates species incapable of long-range migration, and that is one of the virtues of the waif hypothesis.

Among the last common island organisms to be proved capable of distant dispersal were insects. Even on the Hawaiian Islands, with their large human population, there was no way to detect an insect that had just been blown in from California. J. L. Gressitt solved the problem in the 1950s by towing a large fine-mesh net behind an airplane near the islands. It was like the discovery of plankton in the sea a century earlier. Insects and spiders are abundant even high in the air, and the species represent groups in the same proportions as those of the insect faunas of oceanic islands.

2

Plate Tectonics and Islands

The scientific study of oceanic islands began two centuries ago, but several new factors have made them more inviting objects of study. First, almost all volcanic islands have now been dated. Therefore the rates of such phenomena as erosion or subsidence can be measured, whereas before they were speculative. Second, geological history as a whole has benefitted from intensified study with new tools during recent decades. In earlier times, each new fact either tended to undermine old hypotheses or stood alone. Now, the theories of continental drift and plate tectonics provide a framework that becomes stronger as each new fact is riveted into place. Third, widespread observations can be quantitatively related by plate tectonics. Tectonic plates are rigid, so all points on a plate remain in the same configuration as the plate drifts about. Thus, if the speed and direction of drift of a few points or islands can be established, the drift of all others on the same plate can be calculated. A few very careful observations, although scattered, are enough to add a new accuracy and unity to geologic history.

TECTONIC PLATES

The crust of the earth is a spherical shell of rock that consists of a few rigid plates. There are only eleven giant ones at present, and many smaller; two of the giants seem to be in the process of splitting up. These tectonic plates drift about continually, shifting position and jostling each other. Consequently, the boundaries between them are marked by earthquakes. In-

High fault cliffs cutting basaltic lavas mark the Mid-Atlantic Ridge where it passes through Iceland.

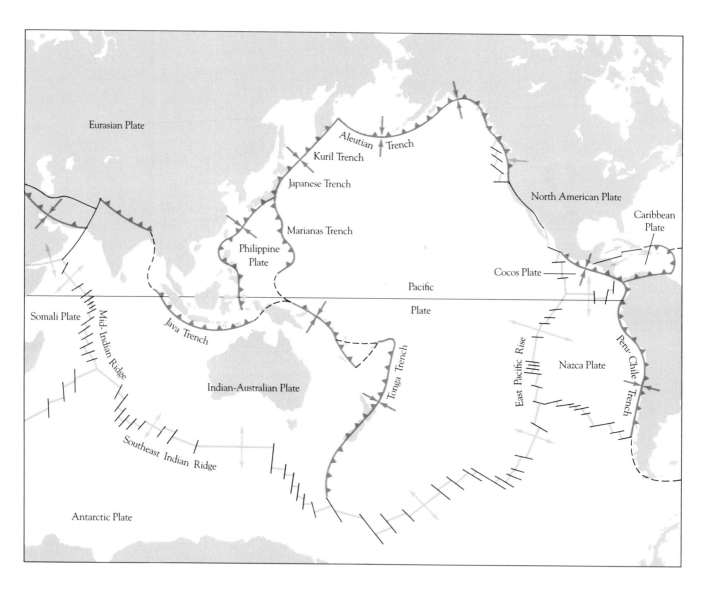

The tectonic plates of the world. Spreading centers are shown in orange, subduction zones in blue. Black lines are transform faults.

deed, one way to define a plate is "a region of the crust lacking earthquakes but ringed by them." If we look at a world map of earthquake epicenters, two things are striking. Almost all the quakes are along the lines of plate boundaries, and few of these lines correspond to the boundaries between continents and ocean basins. Clearly, the forces that move tectonic plates are so mighty that they can hardly tell the difference be-

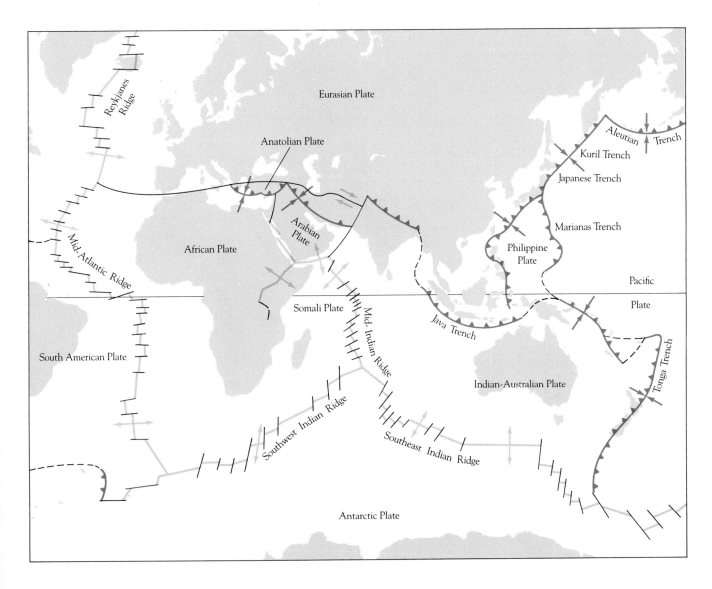

tween high continents and the deep sea floor. Coastal Southern California, for example, is on the Pacific plate and is drifting northwest with the rest of the plate toward Siberia. Eastern California, in contrast, is attached to the North American plate, which extends east to the center of the Atlantic Ocean. Eastern California, with its plate, is drifting slowly to the southwest.

Magnetic anomalies along the Reykjanes Ridge, southwest of Iceland. Colors show areas of magnetic reversal relative to the present orientation. In about 12 Ma, the plates on either side of the ridge have spread about 200 km.

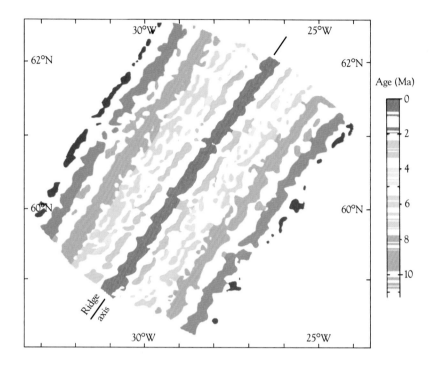

Spreading Centers

Plates are created by solidification of passively upwelling magma, which fills in the cracks where plates drift apart. The cracks are called *spreading centers,* and they are characterized by tensional earthquakes (caused by stretching), which are confined to shallow depths because the hot crust in these places is too weak for stresses to accumulate any deeper. When a spreading center first forms, it may open a crack in either a continent or the sea floor. Such a crack gradually opened between what are now Africa and South America roughly 200 million years ago. Those continents are far apart now, but the seismically active crack still exists at the crest of the Mid-Atlantic Ridge. That ridge is one of a class of great topographic features called, for convenience, "midocean" ridges even though some are nowhere near the middle of an ocean, and the Pacific has sometimes contained two or more of them. Midocean ridges are typically a few kilometers high above the deep ocean floor and, with their sloping flanks, a thousand kilometers wide.

The magnetic field of the earth reverses polarity at intervals on the order of a hundred thousand to a million years. When lava cools, some of the minerals in it act as tiny magnets and orient themselves in the direction of the magnetic field. The spreading crack on the crest of a midocean ridge is frequently filled with lava, which then cools, splits, fills, splits, and so on. Thus the cold rocks of the ridge contain a fairly permanent record, like a magnetic tape recording, of the reversals of the earth's magnetic field through geological time. Indeed, the whole ridge is like a stereo tape recording with magnetic patterns on each flank that are commonly mirror images of each other. The pattern of normal (like now) and reversed magnetic orientations (anomalies) has been dated by comparing rocks of known age on land with those on the sea floor. Inasmuch as most of the magnetic anomalies of the ocean basins have been mapped by ships, the age of most of the vast, deep sea floor is known. Using the width of dated magnetic anomalies, it is possible to measure how rapidly the midocean ridge crest where they were created was spreading apart—even though it was 100 million years ago.

Subduction Zones

The size of the earth has been quite constant for billions of years. Consequently, when a spreading center produces an area of new crust, an equal area of old crust must be removed from the earth's surface somewhere. The opening of the whole Atlantic Ocean basin, for example, resulted in the loss of an equivalent area, mainly in the Pacific. The regions where tectonic plates drift together and crust is lost are mostly *subduction zones*. In such a zone, one plate plunges beneath the other, usually at an angle between 30° and 45°, and goes on down for hundreds of kilometers into the mantle. The plate that plunges is almost always oceanic crust, because continental crust is more buoyant. The path of the plunging plate can be traced by the earthquakes that are generated. Under Japan, for example, where the Pacific plate plunges beneath the Eurasian plate, the earthquakes are shallow; under the Sea of Japan, farther west, the quakes are deeper; and under easternmost Siberia, they are deepest. Typically, subduction zones have the largest and most damaging earthquakes in the world because the rocks there are old, cold, and able to accumulate large strains before breaking.

The great compressive forces in subduction zones deform the crust into deep oceanic trenches and high continental mountains such as the Alps and Himalayas. The reheating of the plunging oceanic crust and sediment causes magma to liquefy at depth. It rises to the surface to form lines of beautiful volcanoes like the Cascade Mountains of Oregon and Washington and including the most beautifully symmetrical of all— Fujiyama in Japan. Like sea-floor spreading, subduction can take place within continents or ocean basins, but in fact it takes place mainly at the boundaries between continents and oceans. The Pacific, unlike the Atlantic, is ringed by subduction zones and the line of fire of active volcanoes. The reason is not that the zones develop at the edges of continents; as at spreading centers, the forces that move plates are much too great to be influenced by the type of crust. What happens is that the buoyant continents drift to subduction zones and stay there like rafts at the edge of a whirlpool in a river.

Transform Faults

The crest of the Mid-Atlantic Ridge is not straight; it is offset just like the Atlantic coasts of Africa and South America and for the same reason. The offset is an abrupt step, and it is caused by *transform faults*, which, like spreading centers and subduction zones, are one of the three basic elements of plate tectonics. A transform fault is, as the name implies, merely a fault, a cut in the earth's crust, running between the other two kinds of tectonic elements. Most transform faults offset the crests of midocean ridges; a few run between a ridge and a subduction zone; even fewer run from one subduction zone to another. The earthquakes on ridge-ridge transforms are quite shallow (1 km to 5 km deep) because the crust there is young and weak. However, where transform faults cut older crust, earthquakes may be 10 km to 20 km deep.

The most famous transform fault is the San Andreas fault, which transects California and destroyed much of San Francisco in 1906. It is of

Transform faults between spreading ridges (color) are marked by steep, rugged topography created as the plates slide past each other.

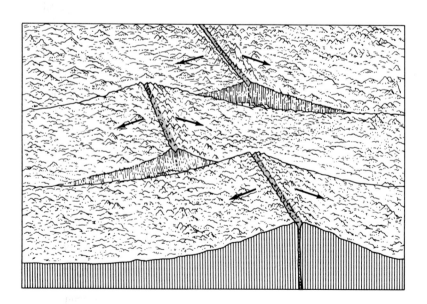

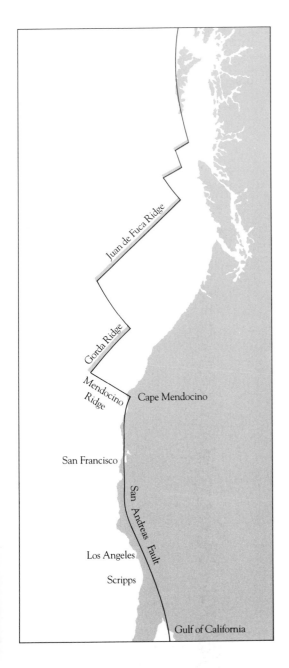

The San Andreas fault is part of the boundary between the Pacific plate and the North American plate.

the most common type, a ridge-ridge transform; thus, its notoriety derives not from unusual geology but from the concentration of people and buildings around it. The San Andreas runs between two "midocean" ridges, one of which actually extends into Mexico and the other of which is only a short distance off Northern California. The southern one is the great East Pacific Rise, which extends from the South Pacific to the mouth of the Gulf of California. This gulf has been created in its present form during the last few million years by the drift of the Pacific plate away from the North American plate. The sea floor of the gulf is broken into a number of short ridges and long ridge-ridge transform faults. The San Andreas fault is one of them; it emerges from the northern end of the gulf and extends into California. Then it passes 100 km or so east of Los Angeles and right through the western part of San Francisco before trending out to sea near Cape Mendocino. The fault ends at the Gorda Ridge, but the plate boundary continues on past Alaska and Japan and ultimately back to the South Pacific. The San Andreas fault, causing centimeters of offset every year, may seem enormous to Californians, but it is only a minor part of the truly enormous perimeter of the Pacific plate.

The faulting along submarine ridge-ridge transforms produces a distinctive topography with long, narrow mountains and deep troughs, high volcanoes and great cliffs. The active faulting that generates earthquakes takes place only between spreading centers. However, the distinctive topography is preserved and drifts away from the ridge crest with the growing plates on both sides. The resulting mountain ranges, called *fracture zones*, are typically from 10 to 100 km wide and may be thousands of kilometers long. If the water in the ocean were removed, the great fracture zones would be readily visible even from the moon and seem so straight and evenly spaced as to appear artificial.

AGING AND SUBSIDENCE OF PLATES

The earth's rigid surface layer, or *lithosphere*, is almost wholly in buoyant equilibrium, or *isostasy*. Because the lithosphere effectively floats on a weak plastic layer, or *asthenosphere*, its elevation is related to its density. High mountains are composed of rocks of low density, and the deep sea floor is composed of rocks of high density.

The crest of a midocean ridge rises high above the deep sea floor because the young crust created at the spreading center is hot. As the crust drifts away, it cools by conduction to the cold sea floor; it grows denser, so it subsides to form the sloping flanks of the ridge. Cross-sectional profiles of ridges show that their flanks are concave upward between the high crest

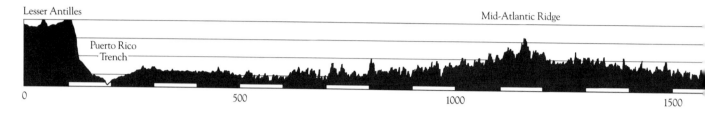

Lesser Antilles   Mid-Atlantic Ridge

Puerto Rico
Trench

0   500   1000   1500

An echo-sounding profile across the Mid-Atlantic Ridge shows endless hills and mountains superimposed on broad concave slopes.

and the deep basins on either side. It is immediately clear that cooling and subsidence are more rapid when the plate is young than later. The relation between depth and age has been determined in thousands of places and is empirically expressed for crust younger than 60 million years (Ma) as

$$d_t = d_0 + Kt^{1/2}$$

where d_0 = initial depth (2500–2600 m)
d_t = depth (in meters) at time t (in Ma)
K = a constant (320–360 m)

In short, the depth increases with the square root of time. The average initial depth and the constant K are still being determined within a narrow range. The heat flow and other properties of plates also vary with $t_{1/2,}$ and all these variations can be explained by simple physical models. Thus, it is possible to calculate the expected depth of the sea floor if its age is known. Likewise, the subsidence history of a plate, its depth at any time in the past, can be calculated. The ability to make these calculations has greatly improved understanding of the elevation and subsidence of islands.

Oceanic crust older than 60 Ma is not known to subside according to the same time relation as younger crust. It has been suggested that heat from the interior of the earth has conducted through the whole plate by that age, so there is no further cooling. The matter is controversial at present, and it is not possible to calculate the thermal history of very old oceanic crust.

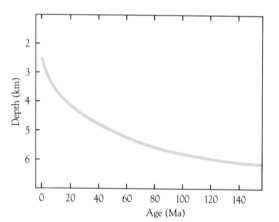

The relation between depth and age of normal oceanic crust.

Thickness and Strength of Plates

Isostasy is commonly achieved by *local support*. For example, a continent or a mountain range may be considered to be floating, buoyed up by the liquid asthenosphere directly below it. However, some support is *regional*, distributed in the area around a mountain. The phenomenon can be visualized in terms of a skater on thin ice. The skater is a load on the ice. If he breaks through and is buoyant, his weight is locally supported. How-

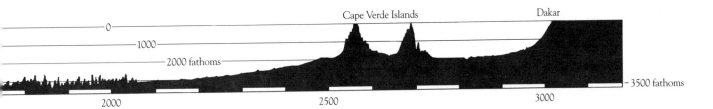

ever, if the ice supports him without breaking, it is pushed downward in a dimple. Although it is not so obvious, the ice is also arched upward in a ring around the dimple—the skater's weight is regionally supported.

Oceanic islands are loads on the lithosphere, and some are supported locally and some regionally. It depends on the age and thickness of the lithosphere when a growing volcano exerts a load.

The top of a plate is the sea floor, whose temperature is about 0°C. The bottom of the plate can be defined in various ways. For example, the top of the asthenosphere, which is in many places about 100 km below the top of the lithosphere, may be taken as the plate's lower boundary, or it can simply be defined as an isotherm—commonly, 1200°C. (The accreting edge of a plate is at a spreading center, where magma is injected at temperatures between 1000°C and 1200°C. Thus, the temperature of a drifting plate is about 1200°C at the side and bottom when the plate is being created.) In any event, the thickness of a plate ranges from 0 km to 100 km, and it increases with age.

The thickness (Z) of a plate down to the 1200°C isotherm varies with age as follows:

$$Z = 9.4t^{1/2} \text{ km}$$

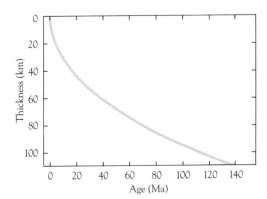

Oceanic crust thickens as it ages.

Thus it thickens rapidly at first but is only about 30 km after 10 Ma, and it does not reach 100 km for more than 100 Ma. The *elastic thickness* is the thickness of the upper layer of the lithosphere that gives regional support to loads. The elastic thickness also varies with $t^{1/2}$, but the constant is much less than 9.4. Although some uncertainty still exists, it appears that the elastic thickness is no more than 10 km at 10 Ma, and may not exceed 40 km at any age. In any event, it is established that large volcanic islands on very young crust, like Iceland or the Galapagos Islands, are locally supported. On the other hand, even large islands like Hawaii do not break through old lithosphere and achieve local support. Instead, the lithosphere deforms like unbroken thin ice, and Hawaii rises from a deep that is ringed by a broad, low arch. Smaller volcanic seamounts and islands have similar but subtler effects on the lithosphere, depending on its age.

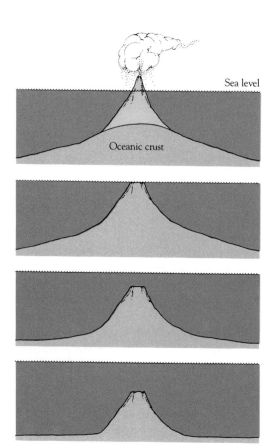

Growth, truncation, and subsidence of a large volcano on a midplate swell.

Midplate Swells

Although most of the sea floor is at a depth appropriate for its age, in some places it is anomalously shallow. The most noteworthy of these places are *midplate swells*, more or less circular or oval areas 500 km to 1000 km in diameter that are commonly 1000 m to 1500 m too shallow in the center. Midplate swells normally underlie volcanic islands; examples are the swells from which rise the Hawaiian, Marquesas, Society, and Samoan islands, in the Pacific, and the Cape Verde Islands, in the Atlantic. Indeed, most active volcanoes within plates are on swells and thus, presumably, most dead volcanoes were on swells when they were active. This speculation is confirmed by the relief of *guyots*, drowned ancient volcanic islands that were eroded down to sea level before they sank beneath the waves. The relief of a truncated island is the distance from the nearby deep sea floor to sea level. This relief is preserved if the truncated island sinks below the sea surface (when it stops being eroded), and thus the relief of a guyot indicates the local water depth when it was truncated by waves. Many guyots have a relief of 3000 m to 4000 m, indicating that they were truncated in water depths found only near the crest of midocean ridges or on midplate swells. It can be established that many guyots are much younger than the crust from which they rise; therefore, if they also have low relief, they were active on midplate swells rather than midocean ridges.

If the date of active volcanism of a guyot is known, and it was truncated rapidly, the date of truncation is known and thus the local water depth at that time. It will be shown later that the duration of truncation depends on the size of an island. However, for guyots with small summit platforms, truncation takes only a few million years, which is often within the margin of error for determining the date of active volcanism. Given the age of a guyot and the present depth of its summit, its average rate of subsidence can be calculated. In many circumstances, the guyot's relief can be taken to be the initial depth of the midplate swell on which the guyot grew. This information can be used to test hypotheses regarding the origin and history of midplate swells.

Three origins have been proposed for midplate swells: rising mantle convection that arches the lithosphere, addition of low-density material to a plate, causing it to rise isostatically, and thermal rejuvenation. The last hypothesis was proposed by Robert Detrick and the late Thomas Crough in 1973. I had noted in 1969 that drilling of atolls in the Marshall Islands in the central Pacific showed that they were sinking at the same rate as younger lithosphere. Detrick and Crough observed that many midplate swells have the depth of standard lithosphere with an age of 25 Ma. From

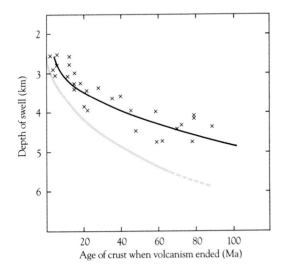

Depth of midplate swells when associated volcanism ceased (black curve) is considerably shallower than normal for crust of the same age (color).

In this painting by Chesley Bonestall, a group of guyots is seen as if the ocean were removed.

the drilling of atolls and the shape of the Hawaiian swell, they calculated that swells subside like 25-Ma lithosphere. They then reached the bold but logical conclusion, in 1978, that the lithosphere of swells has been *thermally rejuvenated*—thinned by heating from below and elevated by isostasy.

Marcia McNutt and I confirmed the value of thermal rejuvenation as an explanation for midplate swells in 1982. We had examined 33 isolated oceanic volcanoes, including twelve guyots. The volcanoes ranged in age from active to 90 Ma and were on lithosphere from 3 Ma to 163 Ma old. The depth of the lithosphere when volcanism ended was less than the standard depth for lithosphere of that age, so the volcanoes were or are active while on midplate swells. A plot of the depth of the swells when volcanism ended versus the age of the lithosphere at that time shows that lithosphere less than 12 Ma old had been uplifted to a depth of 2500 m to 2600 m—the same depth as the crest of midocean ridges. Older lithosphere had been uplifted to a depth that depended on its age. The relation can be expressed easily in terms of thermal rejuvenation—older lithosphere is commonly rejuvenated by about one third of its age. It is not rejuvenation by the fabled Fountain of Youth, but many a sixty-year-old would be happy to be forty again.

The midplate swells under our twelve guyots and the atolls of Enewetak and Midway sink at rates that depend on the age of the lithosphere—the younger the lithosphere, the faster the swells subside. But whatever the age, they subside faster than normal lithosphere of the same age. Once again, the relation can be expressed easily in terms of thermal rejuvenation—regardless of its age, the uplifted lithosphere subsides as though it had the age of normal sea floor with the same depth. A midocean ridge flank with an age of 15 Ma has a standard depth of about 4000 m. Old, deep lithosphere that is uplifted to 4000 m on a midplate swell subsides as though it were 15 Ma old.

DRIFTING PLATES

The fact that tectonic plates are rigid might seem entirely obvious because solid rock, such as Gibraltar, is the very image of rigidity. However, it is all a matter of scale and the duration of stressing. If the whole earth is shocked almost instantaneously by a great earthquake, it rings like a rigid bell at low frequencies appropriate for its size. On the other hand, geologists have known for a century that the very slow application of pressure to heated rock can cause it to deform like toothpaste or soft clay. The solid,

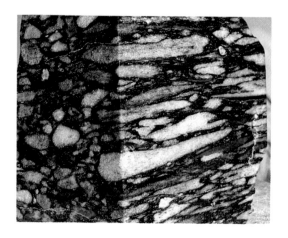

Metamorphically deformed cobbles.

A crane could not lift the earth's crust under Texas because the weak crust would bend and collapse.

spherical cobbles of an ancient beach can be drawn into elongate shapes like pencils by the process of metamorphism. Thus, dealing as they do with millions and billions of years, geologists tend to think of the earth not as rigid but as yielding and plastic.

Moreover, a famous scientific paper in 1937 had demonstrated that it was impossible to lift up a large area of continental crust. (It had a striking cartoon of a giant crane lifting up the earth's crust, 30 km thick, under the state of Texas.) The solid crust proved to be too weak to be lifted at the edges without sagging in the middle. Consequently, it was a considerable surprise to geologists and geophysicists when it was proved that enormous tectonic plates are rigid. Not in the vertical direction—plates do bob up and down locally by small amounts to maintain isostasy, and they cannot be lifted any more than Texas can—but horizontally, they are inflexible.

The rigidity of the plates was demonstrated by appeal to a theorem of the mathematician Leonhard Euler. This states that if one rigid shell moves over another without changing direction, two diametrically opposed points must remain fixed. These points are called *Euler poles*. The motion of any point on the moving shell may be considered as a rotation around an axis that connects the Euler poles. Relative to the inner shell, points on the moving shell traverse circular arcs centered on the Euler poles. If a tectonic plate is rigid, it can be considered a fragment of a spherical shell (the lithosphere) moving over an inner shell (the earth's mantle). Then its movements must conform to Euler's theorem.

If the Euler poles were, by coincidence, the poles of rotation of the earth, the circular arcs would exactly coincide with the parallels of latitude. In fact, they rarely do, so one must visualize *Euler latitudes* measured from the actual location of the Euler poles. The motion of a whole rigid plate can be described accurately only by an angular velocity around an Euler pole. The velocity of a given point, however, may be expressed usefully as a linear rate, usually as millimeters per year. This rate varies with Euler latitude from zero at the poles to a maximum at the Euler equator. The highest spreading rate known is 170 mm/yr, in the southeastern Pacific, but it may have been faster in Cretaceous time.

A point on the side of a drifting rigid plate, like all other points, must follow a circular arc, and thus the transform fault boundaries at the sides of plates must lie on circular arcs. The crust of the North Pacific was demonstrated to be a rigid plate—by analysis of the motion on the faults that bound it—by Dan McKenzie and Robert Parker in 1967. From California to Alaska to Japan, all the faults plotted along circles centered on an Euler pole near Greenland. Jason Morgan, also in 1967, showed that transform faults and fracture zones between plates lie along circular arcs around an Euler pole. Morgan also showed that, in the Atlantic, the widths of mag-

A hypothetical midocean ridge superimposed on the Pacific Ocean. Because rigid plates drift apart in accord with Euler's Theorem, the angular velocities are constant at different Euler latitudes, but the linear rates of relative motion are greatest at the Euler equator and least at the Euler pole.

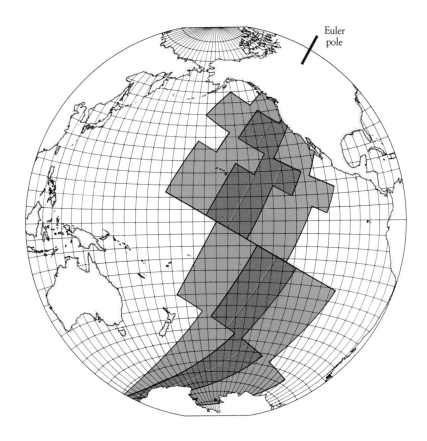

Euler pole

netic anomalies and thus the rates of spreading vary with the Euler latitude in exact correspondence with Euler's theorem.

Relative and Absolute Drifting

Angular rotations around an Euler pole define the motion of one plate relative to another. The local spreading rate indicates how fast two plates are moving apart, but it implies nothing about their motion relative to the earth's rotational poles or equator. Two plates can spread apart, for example, even though both are drifting west, if their rates of drift are different. If plate tectonics is to be useful in reconstructing geological history, it would be desireable to tie plate motion into a normal geographic framework. It is fairly easy to relate plate motion to the equator or poles because both the earth's magnetic field and its climatic zones leave traces in the geological record. Latitude, for example, determines the dip angle of the

This prize-winning (1736) chronometer was the first clock accurate enough to be used to determine longitude.

magnetic field, and the dip is recorded by the orientation of magnetic minerals, not only in volcanic rocks, but also in some kinds of sediment. Likewise, the global wind system influences the location of deserts and the orientation of sand dunes and plumes of volcanic ash.

Latitude is always relatively easy; longitude is the problem. The ancient Greeks could measure differences in latitude, but it was not until the eighteenth century A.D. that the invention of the chronometer permitted the determination of longitude. The difference in difficulty is easy to understand. The earth has a natural north and south pole and equator because it is spinning. Likewise, it is easy for a sailor to measure latitude from the position of the sun and stars relative to the horizon. In contrast, longitude is purely an arbitrary convenience for sailors and geographers. At one time, different western European nations made maps with a zero longitude through their national capitols. The present global acceptance of a zero longitude through the astronomical observatory at Greenwich, England, is a rather recent development. How, then, is there any hope of finding indicators of longitude in the geological record? Surprisingly enough, such indicators have been found. For the history of their discovery, it is necessary, as in most things related to oceanic islands, to go back to Charles Darwin on his five-year voyage on H.M.S. *Beagle*.

Volcanic Age Sequences

Darwin reached Tahiti in November 1835—still eleven months from home. The American geologist James Dwight Dana followed in 1839, on the multiship U.S. Exploring Expedition. Although the naturalist Sir Joseph Banks had seen the Pacific islands much earlier with Captain Cook, Darwin and Dana were the first geologists to do so. Darwin, it should be noted, considered himself at that time to be primarily a geologist. Between them, the two geologists discovered the orientation and age sequence of the Pacific islands that would enable Jason Morgan 130 years later to determine paleolongitudes. Darwin's theory of the origin of atolls will be discussed at length in Chapter 7. Briefly, he conceived that volcanic islands subside and that, in coral seas, the subsidence transforms the reefs that fringe the volcanoes into barrier reefs and later into atolls. Thus, different types of islands represent stages of development. With less certainty, islands can be compared, and an atoll can be taken to be older than an extinct volcano with a barrier reef. Darwin, however, ventured no such comparisons.

Darwin did not publish his theory in detail until his book *The Structure and Distribution of Coral Reefs* came out in 1842. However, he obviously talked about it long before then because Dana read about it in a newspaper in Sydney, Australia, in 1839. Dana was electrified by the

Different stages of erosion on the young island of Hawaii (left) and the older island of Molokai (right).

theory, because he had already come to similar ideas—and he was still in a position to test them on the Exploring Expedition. It was he who would be able to correlate the relative age of islands with their distribution.

From Sydney, the U.S. Exploring Expedition sailed on to the volcanic islands in Samoa and Hawaii. Three months in the latter group gave Dana a remarkable insight. Nothing could be more obvious to a geologist now than an age sequence in the Hawaiian Islands. Flying in from the south, one first sees the smooth carapace of the gigantic active volcanoes of the island of Hawaii. Then past Maui to Oahu and Kauai, the islands are ever more deeply eroded into knife-edge ridges and enormous valleys. Assuming only an equal intensity of erosion, the age sequence is manifest. Dana deduced all this at a time when few scientists even believed that valleys are eroded by the streams that run through them. Dana concluded that depth of erosion "is therefore a mark of time and affords evidence of the most decisive character." Darwin, in the same circumstances, might not have drawn any such conclusion because he thought that most valleys had been eroded by ocean waves. However, his views were derived from field trips in the complex continental rocks of home.

Dana was cautious in extrapolating from what he knew, namely that the order of *extinction* of volcanoes was Kauai, west Oahu, west Maui, east Oahu, northwest Hawaii, southeast Maui and southeast Hawaii. His relative order of extinction of pairs of volcanoes on individual islands was completely correct, because he could see where erosion had exposed the overlap of the younger volcano on the older. The erosional age sequence from island to island differed slightly from the sequence of extinction of individual volcanoes, but Dana thought that perhaps this was so only because minor, local vulcanism continued long after a great pulse of rapid activity built the vast bulk of the shield volcanoes. Dana's ideas on the *commencement* of eruptions were limited because "no facts can be pointed to which render it even probable that Hawaii is of more recent origin than Kauai, although more recent in its latest eruptions." He thought that eruptions might have commenced in early Paleozoic time—perhaps 400 Ma ago. Dana, the discoverer, seems to have been one of the very few scientists to make any distinction between the sequences of origin and extinction of volcanic archipelagoes.

Dana had personally visited the major volcanic archipelagoes of the Pacific and, like any scientist who works with nautical charts for years at sea, he knew that the islands had the same west-northwest trend. Thus, after his analysis of the Hawaiian group, he was prepared to opine that the Society Islands were "first extinct at the northwest end." Dana did not know of the active volcanoes at the southeastern end of the Society Islands because they have not yet grown above the sea surface. However, Mehetia, the island between Tahiti and the active seamounts, has had recent lava flows. Farther to the northwest are a series of deeply eroded islands beginning with Tahiti. The most notable feature of the islands is that the lagoons grow broader as one sails to the northwest. At the northern end of the archipelago, the central volcano disappears, and all that remains is an atoll—a lagoon circled by a reef. Dana's age progression in the Society Islands thus was a side-by-side display of what Darwin visualized as the stages, one above another, in the subsidence of individual islands.

In the Hawaiian Islands, there were reef-circled pinnacles northwest of Kauai, and beyond them were atolls. The age sequences in the Society and Hawaiian islands were in the same direction. Dana found a third sequence from active volcano to atoll in the Samoan group, but it was in the opposite direction. Dana sent Darwin a copy of his *Geology of the U.S. Exploring Expedition* when it was published, in 1849. Darwin responded "last night I ascended the peaks of Tahiti with you. . . . " The whole scientific world was aware of the remarkable age progression of the Pacific islands.

The discovery of Midway atoll, in 1859, was the last to be made on the surface of the sea. However, the vast floor of the sea between the sparse islands was unknown. It remained so until the end of World War II, when advances in anti-submarine warfare yielded superior echo sounders. The 1950s became a golden age of deep-sea exploration, and geologists once again addressed the questions about oceanic islands that had been raised a century before. The Dana age progression was a central fact of island geology, and it might have been expected to dominate hypotheses regarding the origin of Pacific islands, but it did not. The idol of the explorers in the 1950s was Harry Hess, who had discovered the guyots (predicted by Darwin) during the war. Perhaps the reason the Dana progression was ignored at first was because Hess had ignored it in his explanation of the origin of the Hawaiian Islands. He proposed that fissures and tension cracks had opened along a great "transcurrent" fault—one of a class of faults that are typically straight and have only horizontal motion. Inasmuch as the whole length of such a fault is active at once (on a geological time scale), an age progression would hardly be expected and none was mentioned.

As sea-floor exploration progressed in the 1950s, the islands tended to be ignored because so many large undersea volcanoes were discovered. These were of two general types: flat-topped guyots, which were drowned ancient islands, and pointed-topped seamounts, which had never been truncated by waves. The guyots were assumed to be geologically old, at least old enough to grow 4 km to 5 km up from the sea floor, be truncated, and sink to various depths. Moreover, some were dredged and proved to be roughly 100 Ma old. Many of the seamounts and guyots were in lineations trending northwest—like Dana's island chains with age progressions. However, some groups, such as the Emperor Seamounts, had a distinctively different trend that was almost due north. This was particularly intriguing because the purely submarine Emperor trend was clearly an extension of the Hawaiian trend with a connection at a "bend" or "elbow." Moreover, the Hawaiian trend itself seemed to continue even beyond Midway atoll. Evidently, another stage of submergence carried even atolls beneath the waves.

The discovery of the Emperor-Hawaii bend might have made it clear that the lineations were produced by some process that acted sequentially and could change directions. However, these facts were clouded by other discoveries. Some groups of guyots, such as the Mid-Pacific Mountains, were in clusters rather than lineations. Some lineations seemed to overlap rather than meet at a bend. Most confusing of all, in many places, atolls, barrier reefs, high islands, and guyots were all mixed together. The reality of the age progression seemed highly questionable.

In 1957 L. J. Chubb made a considerable advance by simply ignoring the sea-floor discoveries. First he showed that almost all the high islands in the Pacific have west-northwesterly trends. Not just the ones noted by Dana, but four other groups as well. But atolls are relatively senile chains of islands, and Chubb showed that they have a different trend that is more northwesterly. He would have included most of the submarine volcanoes

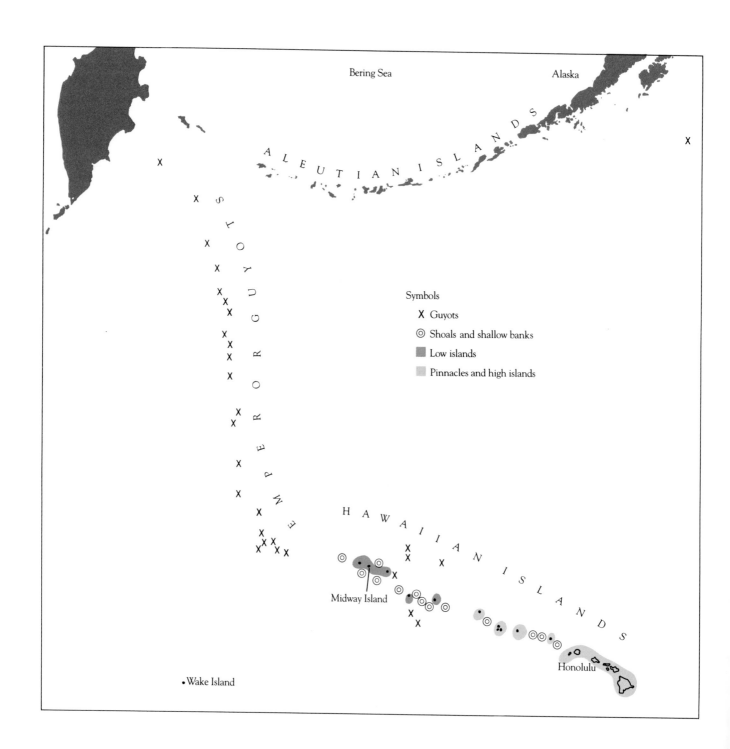

Symbols

X   Guyots

◎   Shoals and shallow banks

⬛   Low islands

⬜   Pinnacles and high islands

in his generalization that the "direction of earth-movements" had changed gradually during geological time from NNW to NW to WNW.

Hot Spots

The nature of these "earth-movements" was proposed by J. Tuzo Wilson in 1963. Harry Hess and Robert Dietz had already proposed that the oceanic crust is created at the crest of midocean ridges and spreads to each side. But Hess called his hypothesis "geopoetry," and Dietz viewed his seminal paper as a "pot boiler"; few people took sea-floor spreading seriously in 1963. Wilson did. He proposed that the sea floor spreads because it is carried along by convection currents that rise from deep in the mantle in thin sheets, flow horizontally beneath the crust, and then return to the mantel in thin sheets. That left a motionless core in each convection cell. Wilson visualized in this core a fixed source of lava, which would rise to build volcanoes. The horizontal convective flow would carry volcanoes away from the source, and the result would be a linear age progression of volcanoes.

But the reality of the age progressions still had to be confirmed. In 1964, Ian McDougall (and later, Brent Dalrymple and others) began to publish the ages of insular volcanic rocks. Dana had been sure only of the sequence in which the volcanoes became *inactive*. The isotope geochemists showed that each giant island in the Pacific was produced in only a million years or so, and, although there was minor volcanism later, the Dana progression indeed gave the sequence of island formation.

In 1972, Jason Morgan, one of the inventors of plate tectonics, applied it to the problem of the lineations of islands. He showed that the island lineations of the Pacific plot as circular arcs around an Euler pole and that the spacing of islands in the age progressions depends on the Euler latitude. However, the islands do not indicate relative motion between rigid plates but relative motion between one rigid plate and a framework of lava sources called *hot spots,* in the mantle. If the mantle itself is considered motionless and hot spots do not move around in it, Morgan had tied the past motion of plates into the present geographical framework.

Lines of volcanoes have come to be called *hot-spot tracks,* and they have been studied intensely all over the world. The tracks on different plates can be compared from a knowledge of the relative motion between plates. It appears that hot spots do indeed lie in a fairly rigid framework imbedded in the mantle. So-called "absolute motion" of plates is relative to this framework. Individual hot spots persist for 10 Ma to 100 Ma or

The Emperor Guyots and the Hawaiian Islands are a single chain of volcanoes formed as the Pacific plate drifted over the Hawaiian hot spot. Presumably, the volcanoes at the bend were active when the plate changed direction.

In this northward view of an imaginary, simplified ocean basin, two plates spread apart at a mid-ocean ridge, which is offset by a large transform fault in the middle distance.

The shiny black lava produced at the spreading ridge is quickly covered by sediment, white from the calcium carbonate shells of dead protozoans, that rains down from the surface layers of the ocean. As the drifting plates cool and thicken, they subside. In water more than about four kilometers deep, calcium carbonate dissolves and the accumulating sediment is a rich brown color.

Both plates happen also to be moving due north, at right angles to the spreading movement. The overall direction of drift can be seen in the lines of volcanoes formed as the plates drift over hot spots fixed in the earth's mantle.

The volcano rising from the hot spot in the foreground is in tropical waters. Like the older, dead volcanoes that arose from this hot spot, it is destined to acquire coral reefs and become an atoll as it drifts off the midplate swell over the hot spot. The other hot spots are in cooler waters. Unprotected by coral, their volcanoes are quickly planed off by erosion and sink beneath the sea as guyots.

At left and right, the oceanic plates plunge into the mantle at subduction trenches. Melted material from the plunging plates rises to form lines of volcanoes parallel to the trenches.

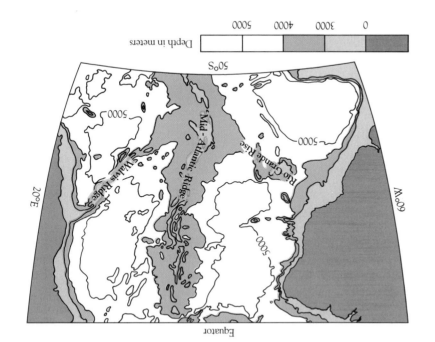

Equator

Depth in meters

0	3000	4000	5000	

V-shaped submarine ridges of the South Atlantic. Many such ridges were once islands.

more, and, if they drift, it is by no more than a few millimeters per year. Their characteristics indicate that they consist of long, narrow plumes of magma rising from the hot lower mantle. These fixed plumes penetrate the lithosphere and rapidly build volcanoes, which become extinct when drifting separates them from the source area.

Morgan's analysis included the origin of the Emperor-Hawaii bend. Inasmuch as the hot spots are fixed, the Pacific plate simply changed its direction of motion without interrupting the production of volcanoes. From other hot-spot tracks of the same age as the Emperor trend, an older Euler pole was established and with it the motion of all such tracks for about 80 Ma in the Pacific.

Hot-spot Tracks and Plate Boundaries

Hot-spot tracks were first discovered in the interior of a drifting plate. However, plate boundaries also drift. What happens when a spreading center or a transform fault passes over a hot spot? Iceland sits astride the Mid-Atlantic Ridge, which is spreading apart slowly. Voluminous flows from numerous volcanoes keep the opening rift filled, but what if volcanism ceased? The island would split and the separate halves would drift away on their respective plates. Exactly this phenomenon has occurred

many times. In the South Atlantic, for example, are two submarine volcanic ridges making a V, with the base on the crest of the Mid-Atlantic
Ridge. The V was generated because the midocean ridge crest was over a
hot spot but each flanking plate was drifting north as well as spreading east
or west. The American plate, in short, drifted northwest, and the African
plate drifted northeast. Very detailed studies of spreading centers by means
of research submersibles have found tiny volcanoes that have split apart.
Likewise, regional mapping discloses pairs of volcanoes that are symmetrical around a spreading center. Thus, this phenomenon of splitting volcanoes is a commonplace—as might be expected, considering the coincidence of spreading and volcanism that is required to generate the oceanic
crust.

A rarer phenomenon is the drifting of an active transform fault over a
hot spot, but examples may exist. Visualize what would happen if an
east-west ridge-ridge transform fault were to drift toward the north over
the Hawaiian hot spot. The present line of volcanoes on the western plate
would continue to drift toward the northwest and still point toward the
hot spot. However, the volcanoes that were subsequently built on the
eastern plate would drift toward the northeast away from the hot spot. The
volcanoes would form a fragmented V with two sections missing. Something of the sort apparently happened in both the South Atlantic and the
South Pacific but the geological histories are not yet firmly established. In
any event, the simplicity of the theory of plate tectonics makes it possible
to predict the consequences of the intersection of a plate boundary and a
hot spot.

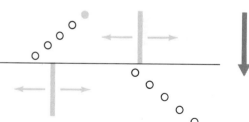

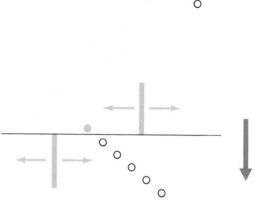

A ridge-ridge transform fault that drifts across a hot spot
should cause a change in the trend of the hot spot trace.

3

Distribution of Oceanic Volcanoes

Charles Darwin, in his *Structure and Distribution of Coral Reefs,* published in 1842, was the first to show the value of maps that identified different types of oceanic islands. By distinguishing among atolls, barrier-reef islands, high islands, and active volcanoes, he was able to show that subsidence was characteristic of some regions and elevation of others. He could also show the relation of active volcanoes to these regions. However, geographical and geological knowledge were still primitive at the time, and he emphasized the value of continuing to map the distribution of types of islands. Since then, the most striking addition to the geography of the oceans has been the discovery of undersea mountains, ridges, and plateaus. Far more numerous than islands, they differ only because their tops remain (or have sunk) below sea level. Like most islands, they are of volcanic origin. In terms of modern theories of global tectonics, lithospheric plates drift over hot spots, relatively fixed plumes of hot rock in the mantle, and these plumes are the major sources of oceanic volcanoes. If this model is correct, the duration and rate of discharge of the plumes, as well as the vulnerability of the plates to penetration by the plumes, are the factors that should dominate the distribution of oceanic volcanoes, and even their size, shape, and arrangement in groups.

VOLCANISM

Volcanism is of five major types: island-arc volcanoes, continental flood basalts, abyssal flood basalts, spreading centers, and ocean-basin volcanoes. Only the first two and the last were known in Darwin's time, and

An oceanic volcano grows into the air in the Galapagos Islands.

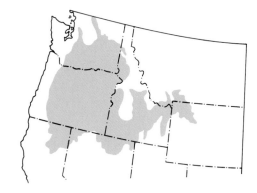

Flood basalts (color) buried much of the northwestern United States in Tertiary time.

until after World War II they still dominated thinking about volcanism. It is only in the last thirty years that it has been realized that volcanism is predominantly a submarine rather than a continental phenomenon. This can be seen by comparing the volume of lava, ash, and associated shallow intrusive rocks generated by the different types of volcanism.

The subduction of lithospheric plates produces very long lines of regularly spaced volcanoes parallel to oceanic trenches. At sea, the volcanoes form lines of remarkably regular conical islands that extend over the horizon in a regular geometrical perspective. Just as midocean ridges are not necessarily found in the middle of oceans, island-arc volcanoes are not all found in the sea. In fact, most continental volcanoes are of the island-arc type—the volcanoes of the Cascade Mountains, for example. The lavas generated by subduction are relatively viscous, and thus volcanoes in island arcs are notorious for devastating explosions. The famous eruption of Krakatou (Krakatoa) in 1883 was in an island arc. It hurled roughly 10 km^3 of material into the upper atmosphere, and the ash cloud drifted around the world several times. Even it was not in a class with other island-arc volcanoes, such as Tambora, which ejected 100 km^3 of ash in 1815, or the Indonesian volcano Toba, which erupted 1000 km^3 of lava flows about 75,000 years ago. Considering the hundreds of volcanoes in island arcs, and the enormous volume of a few spectacular eruptions, it may be surprising that only about 1 km^3 of liquid rock is thought to be emitted in island arcs each year.

Continental flood basalts fortunately have never erupted in historical times. What might happen if they did can be visualized from the detailed geological mapping of lava flows that spread out, presumably in only a few days, in the northwestern United States 15 Ma ago. Single flows flooded 1000-km^2 areas to an average depth of 700 m. In geologically brief periods, such flows reached a total volume of 500,000 km^3 in the northwestern United States and 600,000 km^3 in the Deccan region of India. These basalts tend to form enormous plateaus and are readily recognized in the geological record. A general appraisal indicates that their total volume on land is about 10,000,000 km^3. However, they were spread over several billion years, and the average rate of accumulation is only a small fraction of a cubic kilometer per year.

Abyssal flood basalts were sampled in the western central Pacific during the remarkably successful Deep Sea Drilling Project, in the 1970s. The drill penetrated 500 m without reaching bottom. The flood basalts consist of flows on the sea bottom and sills that were intruded under thin sediment. In fact, considering the high pressure at the deep-sea floor and the low density of the sediment, there is little significant difference between sills and flows. Thus the abyssal flood basalts are very similar to those on

Archipelagic aprons (color) in the western Pacific appear to be abyssal flood basalts of Cretaceous age.

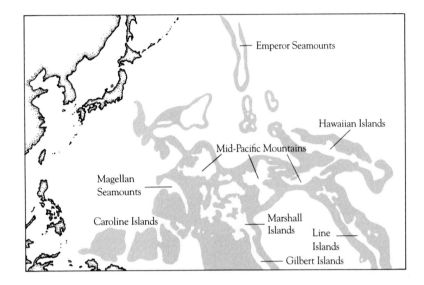

land with regard to mode of emplacement. Drilling has not determined the area of the abyssal basalts. However, their area and volume can be estimated on other grounds. The processes that act at ridge crests always generate linear abyssal hills, whose relief is inversely related to spreading rate. The rapid spreading typical of the Pacific, for example, produces hills with a relief of 200 m to 300 m. Inasmuch as the whole sea floor is a product of spreading centers, abyssal hills should be visible everywhere unless they are buried. Around many western Pacific archipelagoes, the abyssal hills are buried by volumes of material, called *archipelagic aprons*, that seem to imply abyssal volcanism. The thickness of volcanic rock in regions with archipelagic aprons can be measured by seismic techniques; it ranges up to 5.5 km in some places. On average, it is about 1 km thicker than where abyssal hills are not buried in the Pacific. Thin fluid lava flows have been found on ridge crests by research submarines. Thus, the distinctive feature of archipelagic aprons is their enormous volume rather than the type of flow.

From all available data, it appears that the volume of abyssal flood basalts in the western Pacific alone is very roughly 20,000,000 km³, perhaps twice as great as all continental basalts and twenty times as great as any single one ashore. The abyssal basalts poured out on and just under the sea floor at intervals between 70 Ma and 115 Ma ago, in the Cretaceous Period; the continental total accumulated over billions of years. Thus, the abyssal accumulation is equivalent to piling one continental flood basalt

The small craters along Laki fissure in Iceland. In 1783, voluminous lava flows from this fissure buried the surrounding countryside.

on top of another in a small fraction of geological time. It appears that the basalt flooding in the western Pacific may be by far the most voluminous midplate volcanism in the available geological record. Furthermore, most of the sea floor that ever existed has been subducted or accreted to continents, so equally voluminous outpourings of abyssal basalts may have occurred many times in the past but disappeared without a trace.

In contrast to most estimates, the outpouring of volcanic rock and shallow intrusions in spreading centers can be estimated with some confidence because plate tectonics permits reliable interpolations between measurements of spreading rate and crustal thickness. At present, spreading centers are generating 5 km^3 to 6 km^3 per year of volcanic crust in flows and dikes. They have been doing so at least for tens of millions of years and possibly for all of geological time. This average rate is even faster than the average rate in the western Pacific during the episode of Cretaceous basaltic flooding. However the spreading takes place along the enormously long crest of the midocean ridges and thus is not so intense regionally. Some idea of the local intensity of volcanism in spreading centers may be obtained from Iceland. In 1783, the eruption of Lakagigar produced 12.3 km^3, and a single fissure emitted about 10 km^3 to cover 370 km^2 in 50 days. The discharge of 5000 m^3 of lava per second measured at one time was about twice as great as the flow of the Rhine River near its mouth.

The last type of volcanism produces oceanic, or ocean-basin, volcanoes, some of which grow large enough to become islands. Compared with island-arc volcanoes, ocean-basin volcanoes are typically less explosive, less abundant, and much larger. Moreover, almost all active oceanic volcanoes are in only a few sites or groups, in contrast to the long lines in island arcs. Iceland, Hawaii, the Azores, the Canaries, and the Galapagos all include several active volcanoes, but most other active volcanic islands are dispersed.

Sizes and Discharge Rates of Oceanic Volcanoes

Ocean-basin volcanoes have an enormous range in height, from seamounts a few hundred meters high to Mauna Kea, on Hawaii, which rises about 9000 m higher than the surrounding sea floor and has the greatest relief of any mountain in the world. The distribution of sizes has been estimated by studying echo-sounder profiles of seamounts and by counting the number of very large seamounts and volcanic islands of different sizes. The distribution is exponential and remarkably regular. In the younger parts of the Pacific, each area of 10^6 km has, on average, 21 volcanoes with 2000 m of relief, 286 with 1000 m or more, and 1064 with at least 500 m. If these figures are extrapolated to the whole world, it appears that

Left: This volcano in the Galapagos Islands has the typical "shield" or "inverted soup-bowl" shape of ocean-basin volcanoes. *Right:* The tiny volcanic cone of Timakula is typical of islands in arcs parallel to oceanic trenches.

there are roughly 300,000 oceanic volcanoes (active and inactive) with a relief of 500 m or more. Thus, there are incomparably more volcanoes in the ocean basins than on the continents (only a few thousand). However, this fact is largely a consequence of the difference in erosion above and below sea level. Continental and insular volcanoes are eroded away in a few million years, whereas those under the sea are preserved as long as the sea floor on which they stand. If we consider discharge instead of number, a very different picture regarding the intensity of volcanism emerges. The total volume of undersea volcanoes is very roughly 10^7 km^3. Taking the average age of the sea floor to be 60 Ma gives a discharge of 116 km^3/yr or only a sixth of the rate for island-arc volcanoes.

Countless small volcanoes are being discovered by new side-scanning sonar systems.

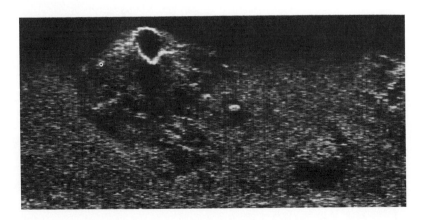

Oceanic islands are the tops of volcanoes that are enormously larger than the famous volcanoes of the continents.

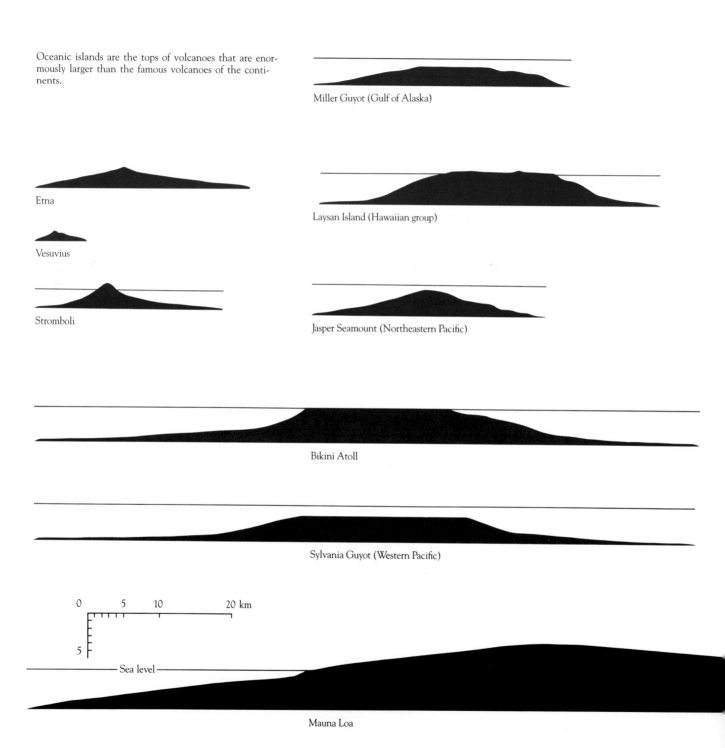

Miller Guyot (Gulf of Alaska)

Etna

Vesuvius

Stromboli

Laysan Island (Hawaiian group)

Jasper Seamount (Northeastern Pacific)

Bikini Atoll

Sylvania Guyot (Western Pacific)

0   5   10   20 km

5

Sea level

Mauna Loa

Because the sea floor sinks as it cools, the rate of lava discharge required to form an island, instead of just a seamount, increases with the age of the crust.

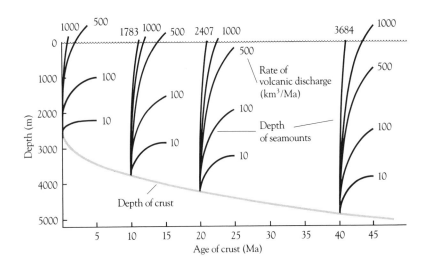

Inasmuch as the relation between subsidence and crustal age is known, a few simplifying assumptions make it possible to calculate the discharge rates necessary to grow a seamount high enough to be an island. Assume that the sides of undersea volcanoes have the 10° slope typical of small seamounts but steeper than the lower flanks of large seamounts. Assume further that the load of the growing volcanoes does not depress the crust and that the density of the volcanoes is typical of solid basalt. In fact, the volcanoes do tend to depress the crust but their density is less than indicated, so these two assumptions cancel each other for these calculations. Most oceanic volcanoes are active for only a few million years, although in the Canary Islands successive volcanoes are superimposed. With constant slopes, the height of a volcano increases only as the cube root of its volume, so it takes a much larger volume (about eight times) to build a volcanic island on 50 Ma crust 5000 m deep than on a ridge crest. On the other hand, very young crust sinks more rapidly than old, and, to become an island, a volcano must grow rapidly enough to compensate for subsidence. For example, a discharge of 10 km³/Ma barely keeps the top of a volcano born at a ridge crest at a constant depth as it drifts away on a subsiding flank. A discharge of 100 km³/Ma will keep such a seamount growing upward, but not fast enough even to reach sea level. If the maximum duration of activity is taken to be 5 Ma, it takes a discharge of roughly 500 km^3/Ma to form an island on young crust and twice that on crust older than 20 Ma. However, the Hawaiian volcanoes commonly have completed the shield-building stage of copious volcanism in only 1 Ma. To become an island in that time on crust 20 Ma or 40 Ma old requires discharges of about 2400 km^3/Ma and 3700 km^3/Ma. Considering that such large volcanoes have very gently sloping sides and large roots, it appears that discharges in the range of 5000 km^3/Ma to 10,000 km^3/Ma, or 0.005 km^3/yr to 0.01 km^3/yr, are required to produce islands in water depths typical of oceanic crust. Presumably, the many isolated guyots that barely became islands in the northeastern Pacific either had discharges in this range for about 1 Ma or greater ones for shorter periods.

The great plumes that produce archipelagoes of large islands flow much more copiously than the minimum for building islands. Hawaii and Iceland are two of the most voluminous islands produced by hot spots. The discharge rates that produced them can be determined in various ways for various time scales. For example, the discharge of the single volcano Kilauea on Hawaii averaged 9×10^6 m^3 per month while it was erupting during a 20-year period. However, it was not always erupting, and the average discharge was only 0.01 km^3/yr. Hawaii has several active volcanoes and their total discharge for a century averaged 0.04 km^3/yr, judged by the measured volume of historical lava flows. The island is famous for the size and frequence of its lava flows, so this historical surface discharge seems large compared with that of other centers of volcanism. Nevertheless, it is grossly misleading regarding the discharge of the mantle plume that has built the Hawaiian Islands. That discharge can be estimated by dividing the volume of the island of Hawaii by its age of about 0.5 Ma. The volume of the island as a topographic feature rising above the sea floor indicates a minimal discharge of 0.22 km^3/yr, or five times the surficial flow rate. The actual discharge is certainly even greater because the island has a great volcanic root that helps to support it isostatically—seismic techniques indicate that the root is as voluminous as the island. Thus, the discharge of the Hawaiian hot spot for the past half-million years has been 0.44 km^3/yr, or ten times the historical rate of flow. The immensity of this discharge can be appreciated by comparing it with the total of 1 km^3/yr emitted by more than a hundred active volcanoes in subduction zones.

The reason for the difference between the historical discharge and the geological estimate is not known. As we shall see, the morphology of the Hawaiian Islands from Hawaii to Midway shows that the plume discharge has varied markedly during periods of millions of years. It is possible, although unlikely, that the plume discharge is decreasing rapidly at present as a part of a long-term change. It is also possible that the historical discharge is only a short-term fluctuation below a long-term average. If so, the normal volcanicity of Hawaii would be ten times as intense as now. Both Kilauea and Mauna Loa would have to discharge large flows almost continuously. No such phenomenon is known anywhere in the world. An alternative is that much of the discharge of the plume is in the form of intrusions within the root and the volcanoes rather than in extrusive flows and ejecta.

Iceland is presently unique in that it is the product of the intersection of a hot spot and a spreading center. It is famous, like Hawaii, for the frequency and volume of its volcanic eruptions, but also for the emission of sheets of highly fluid lava from long rifts caused by spreading. Iceland is an enormous pile of volcanic rock, with an area far larger than Hawaii and a thickness of about 10 km. As in Hawaii, the discharge of volcanic rock can be estimated in two ways. Historically, the measured volumes of flows and ejecta from 1880 to 1980 gives 0.13 km^3/yr as the discharge. Judged from the volume and age of the island, the discharge has averaged 0.06 km^3/yr for 16 Ma. It is apparent that Iceland does not have five times the volume of Hawaii because of more rapid discharge of lava but because of a longer period of accumulation. Iceland would not be so impressive if its plume were under the deep, rapidly drifting Pacific plate instead of the shallow, relatively motionless crest of the Mid-Atlantic Ridge.

Plate Vulnerability

If a volume of the upper mantle is heated enough, it will partly or wholly liquefy into magma, which will rise to the base of the overlying lithosphere. If the volume is small, it tends to solidify as it chills in attempting to penetrate the lithosphere, and it forms intrusive bodies. If the volume is part of a great mantle plume, it will penetrate and build an Iceland or a Hawaii. Intermediate volumes of magma have an uncertain future, but it seems apparent that the chance of penetration decreases in some way related to the thickness of a plate and its rate of drift. Ian Gass and his colleagues formalized the relation in 1978 and defined the vulnerability V in terms of the thickness l and the drift speed u in this way:

$$V = \frac{K}{lu^{1/2}}$$

(K is a constant that must be determined empirically.) Thus the vulnerability is the same if the thickness is 40 km and the drift rate is 1 cm/yr as it is if the lithosphere is only 10 km thick and drifting at a very fast 16 cm/yr. Gass and his colleagues found that large volcanoes have a distribution in reasonable agreement with areas of high vulnerability. Others have questioned the closeness of the agreement; perhaps it is imperfect because most larger plumes can penetrate the lithosphere anywhere, even at the normal maximum of thickness and drift speed. For this reason it seems more desirable to test the hypothesis by studying the distribution of smaller volcanoes.

The distribution of the smallest oceanic volcanoes, 300 m to 1000 m high, is related to the thickness of the lithosphere. This can be demonstrated by determining the number of such small seamounts per unit area of lithosphere of different ages. The age of an area of the lithosphere is known from its magnetic anomalies, but that of individual small seamounts is not, except that they are no older than the crust. Consequently, any individual small seamount, if old, might have penetrated thin lithosphere when a plate was young or, if young, might have penetrated old, thick lithosphere. Fortunately, the average age of individual small seamounts can be inferred from the distribution of a large number. The seamounts on crust less than 5 Ma old are not quite as concentrated per unit area as those on crust 5 Ma to 10 Ma old. On the other hand, the concentration does not increase significantly on crust older than 10 Ma. The only reasonable inference is that most small volcanoes grow on crust younger than 5 Ma and that few, if any, grow on crust older than 10 Ma. The thickness of the crust increases rapidly with age, so the distribution is closely related to the predicted vulnerability of a plate.

At present, the general distribution of medium-sized seamounts as a function of crustal age is not known well enough to study the influence of plate vulnerability. However, there are special situations in which the influence can at least be inferred. One of these is in the northeastern Pacific, where the sea-floor magnetic anomalies are offset 640 km on the Murray Fracture Zone. Because of the enormous offset, the age and thus the lithospheric thickness is much greater on the north than the south side of the fracture zone. On the south side is an abyssal physiographic province called the Baja California Seamount Province because it contains many seamounts 3 km to 4 km high. Being right off Scripps Institution of Oceanography, it was one of the first abyssal physiographic provinces to be discovered, in the 1950s. At the time, it seemed to have more seamounts than most other regions in the Pacific—thus its name. Further exploration has showed that medium-sized seamounts are at least as concentrated in many other regions, so the name is no longer very apt. Nonetheless, it is highly appropriate compared with the Deep Plain Province north of the Murray Fracture Zone. The seamounts of the Baja California Seamount Province tend to lie along short lines that are arcs of circles around the Euler pole for movement of the Pacific plate. In other words, the short

A computer-simulated view of the giant fracture zones of the northeastern Pacific. The relatively smooth floor of the Deep Plain Province (darkest blue) contrasts with the mountainous Baja California Seamount Province to the south, across the Murray Fracture Zone.

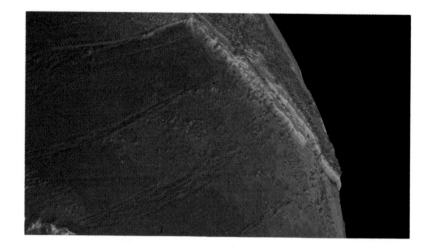

The volcanoes of the Canary Islands rise individually from deep water.

lines of seamounts are products of small, intermittent, but fixed hot spots. The Deep Plain Province and the Murray Fracture Zone have drifted diagonally across the sites of these same hot spots, but the province has hardly any seamounts. Perhaps the hot spots all turned on by chance as the fracture zone drifted diagonally over them. It is more probable that the hot spots already existed when the old, thick lithosphere of the Deep Plain drifted over them, but that they were unable to penetrate it.

Another type of relatively sharp boundary between different lithospheric thicknesses develops as a consequence of thermal rejuvenation. A large midplate hot spot commonly produces active volcanoes, and a swell and thins the lithosphere. Even after the lithosphere drifts away from the hot spot and the primary volcanoes become extinct, the thin lithosphere is still much more penetrable than the unrejuvenated lithosphere that surrounds it. Thus, volcanoes of very different ages should tend to cluster together on old crust, and the clusters should tend to be separated by regions without volcanoes. This is a good approximation of the actual distribution on the very old crust of the western Pacific, where guyots and atolls of very different ages are clustered. It is also a good approximation of the distribution in the South Pacific, where active volcanoes are clustered with atolls and guyots. This phenomenon is carried to the extreme in the Canary Islands, which are on the almost motionless African plate. Eruptions from closely spaced volcanoes have overlapped on some of these islands for more than 15 Ma, while no volcanoes have formed in the spaces between islands.

THE SHAPE OF ISLAND GROUPS

Archipelagoes include islands of various sizes, which are in lines or clusters, and which overlap or are spaced apart. All of these patterns are readily explained in terms of plume discharge, plate thickness, and drift speed. If a plate drifts rapidly, a group is linear like the Hawaiian and Marquesan groups. If a plate is motionless, groups are clustered like the Canary, Azores, and Cape Verde islands. If a plate is thick and drifting, islands tend to be spaced apart like the northern Hawaiian, Society, Samoan, and Marquesas islands. This is thought to be because a plume establishes a conduit through the lithosphere through which it feeds magma to build a volcano. Initially the conduit is vertical, but as a plate continues to drift, the conduit between the plume and volcano is tilted in the direction of drift. Eventually, tilt becomes too much and a new vertical conduit develops over the plume, leaving a space between volcanoes. If some angle of tilt is critical, the thicker a plate the greater will be the spacing between volcanoes, if other things are equal; moreover, fast drift will tend to shorten the life and hence the size of volcanoes.

Plume discharge is also important. The southern Hawaiian Islands, unlike the older ones to the north, rise from a continuous ridge. The thickness and drift of the Pacific plate have not varied as the gross morphology of the Hawaiian Islands has changed, so the ridge must have been caused by an increase in plume discharge through the plate. There are many similar but inactive volcanic ridges in the ocean basins—It clearly is not uncommon for volcanoes to be so closely spaced that their bases overlap more or less completely. If a plate is drifting, a linear ridge is formed. If a plate is relatively motionless, the result is a roughly circular volcanic plateau.

Volcanic Ridges

Easter Island and Tristan da Cunha are young volcanoes of modest size that are the latest products of long-persistent hot spots. At present each island is near a spreading center. The hot spots and spreading centers do not obviously overlap now, but they once did. This is evident from the fact that each of the hot spots has produced two volcanic ridges, on opposite sides of the spreading center, in the shape of a V. From 100 Ma to 60 Ma ago, the Tristan da Cunha hot spot coincided with the crest of the Mid-Atlantic Ridge and produced the Walvis and Rio Grande ridges, which were capped by enormous islands (see the map on page 48). For the last 40 Ma, the hot spot and spreading center have been separated, and the volcanoes generated have been small and separate. Almost exactly the

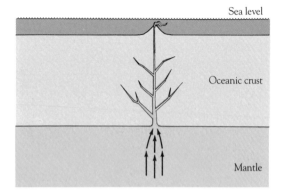

Sea level

Oceanic crust

Mantle

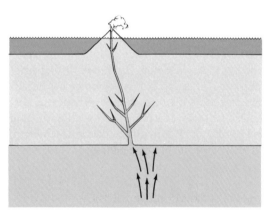

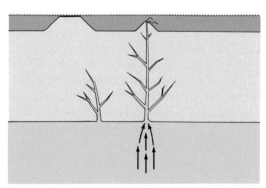

A fixed mantle plume interacts with a drifting lithosphere.

same sequence of events occurred in the South Pacific, where the Easter hot spot simultaneously generated the Nazca Ridge, capped by large guyots, and the eastern ridge in the Tuamotus, which is capped by atolls. It appears that the volcanic discharge of each of the plumes was at least 100 times greater when coincident with a spreading center than when off to one side.

To explain this phenomenon, it is necessary to consider the interactions of plumes and spreading centers. The composition of rocks derived from a plume and from a normal spreading center is different, so the origin of magmas where two possible sources coincide can be determined. The section of the Mid-Atlantic Ridge to the south of Iceland is called the Reykjanes Ridge. It is distinguished because the crest slopes upward toward Iceland and is much shallower than a normal ridge. Chemical analysis of numerous dredge hauls by Jean-Guy Schilling show that the rocks contain a mixture of plume and ridge components and that the plume fraction decreases from Iceland toward the south. Clearly, the Iceland plume spreads laterally along the ridge crest. In the South Atlantic, the plume-generated islands are 50 km to 200 km away from the ridge crest, but Schilling finds that the parts of the crest nearest the islands are unusually shallow and their rocks are partly derived from plumes. So plumes may spread laterally into spreading centers. In the South Atlantic, islands grow over the plumes, but part of the plumes also flows to a nearby spreading center. Jason Morgan has shown that, in the Galapagos Islands and elsewhere, islands are produced not over the plume but where the laterally spreading plume intersects a spreading center. In the Indian Ocean, the Reunion plume may have simultaneously produced both Mauritius overhead and Rodriguez at the distant crest of the Mid-Indian Ridge.

The lateral flow of plumes may partly explain why a hot spot might produce bigger islands when distant from a ridge crest than when near one. However it does not seem to explain the production of the conspicuously voluminous volcanic ridges when the plume coincides with a spreading center, because part of the plume's discharge must then be lost to lateral flow along the midocean ridge. One possible explanation of volcanic ridges is based on plate vulnerability. At a spreading center, there is no impediment to a plume; the entire discharge is available to form volcanoes and contribute to the volcanic layer of the crust. When a midplate plume has to penetrate the lithosphere, it loses part of its discharge to intrusions. However, the plumes that recently produced the isolated islands Tristan da Cunha and Easter Island are now under young, relatively thin crust, and it is unlikely that it could absorb almost all of the plumes' production in intrusions. So some other factor caused these two plumes to produce great ridges in the past.

All else failing, the only remaining explanation seems to be that volcanic ridges contain a large fraction of midocean ridge basalt mixed in with the discharge from plumes. If so, while plumes are contributing to coincident spreading centers, midocean ridge basalts are contributing to volcanic ridges. For some reason, the two sources yield much more together than they do separately.

Leaky Transforms

Some volcanic ridges capped with guyots are so straight that it is difficult to believe that they are hot spot traces. An example is the ridge that lies along the eastern part of the Mendocino fracture zone. It is not flat-topped like an ordinary guyot, but rounded basalt pebbles found there show that the top once extended above sea level as a long thin island. The origin of the ridge is uncertain but it appears to lie along a "leaky transform," a transform fault with a minor component of spreading. The Necker Ridge, west of the Hawaiian group, appears to be an identical feature. The growth of volcanic islands along a lengthy leaky transform might be spectacular— rather like all the Hawaiian Islands erupting at once.

The longest exceptionally straight volcanic ridge is the guyot-topped Ninety East Ridge in the Indian Ocean, and it is a puzzling feature. It appears to lie along one of a group of parallel fracture zones and thus could be a leaky transform, but the guyots are dated by drilling and rather than being contemporaneous they are in an age sequence that is younger toward the south. A complex history is indicated.

The Ninety East Ridge, in the center of this picture, extends almost exactly due north-south for thousands of kilometers.

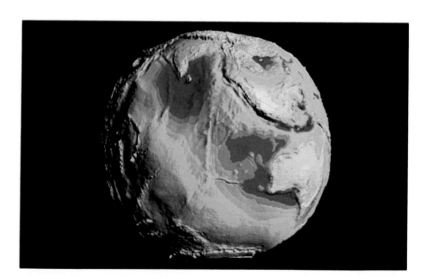

Distribution of types of oceanic islands on the major plates. Iceland is excluded because it is on a ridge crest. Numbers in parentheses are adjusted as if all plates had the same area.

Plate	Average speed (cm/yr)	Relative area	Active volcanoes	High islands	Banks	Atolls and guyots	Total
American	2.8	1	0	5	16	12	33
African	1	1	13	30	32	9	84
Indian	7	1	0	2	20	55	77
East Pacific*	6	0.33	7(21)	10(30)	2(6)	10(30)	29(87)
Pacific	10	2	10(5)	43(22)	71(36)	596(298)	659(330)

*The East Pacific is a region made up of the Cocos and Nazca plates.

DISTRIBUTION OF ISLANDS AND PLATE VULNERABILITY

The global, including continental, distribution of active volcanoes indicates an inverse correlation with the product of plate thickness and the square root of the speed of plate drift. We now have maps of modern and ancient volcanic islands and can see whether their distribution is correlated in the same way. Consider first the distribution related to an average speed of drifting as shown in the table on this page. There are indeed many active volcanoes on the slowly moving African plate, but also on the rapidly drifting plate of the Pacific. The relatively young but inactive high islands are distributed in the same way, with many on the plates that have abundant active volcanoes and few on the American and Indian plates, which lack active midplate volcanoes.

The distribution of older volcanic islands differs greatly from that of active and high ones. For example, there are almost as many oceanic banks on the American and Indian plates as on the African one despite the great difference in the number of active volcanoes. As to the more ancient guyots and atolls, the Indian plate has six times as many as the African plate. Moreover, the Pacific plate, which has about the same number of active volcanoes and high islands as the American plate, has twice as many banks and 65 times as many guyots. Clearly, past vol-

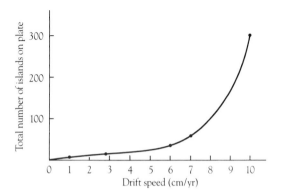

Number of drowned ancient islands on lithospheric plates drifting at different rates (from the table on page 65). Atolls and guyots are most abundant on rapidly drifting plates.

canicity is not related to the present drift speed as modern volcanicity is. It may be argued that past drift speeds differed from present ones, but in fact they seem to have been rather constant for long periods in most parts of plates. If drift speed has not changed, then the number of volcanoes per unit area does not decrease with the square root of the plate speed. Instead, the number increases rapidly—roughly with the cube of the plate speed. However, plate vulnerability logically (as well as empirically, for modern volcanoes) is related to plate thickness as well as to drift speed. Did thickness have the same relation to volcanism and plate vulnerability in the past as it does now?

The Pacific plate is certainly the one of greatest interest with regard to volcanism. Although its area is less than half of the world ocean, it has 25 percent of the active oceanic volcanoes, 45 percent of the high inactive volcanic islands, 60 percent of the banks, and 80 percent of the guyots and atolls. The chief characteristic of the plate for present purposes is that most of it has been drifting at about 10 cm/yr for the past 40 Ma and probably for the past 80 Ma. To a first approximation, it can be assumed that all the volcanic islands in the Pacific grew on crust drifting at the same speed and that the only variable affecting the plate's vulnerability is its thickness.

The table below indicates the remarkable fact that active volcanoes and youthful high islands are at least as abundant on crust 50 Ma to 100 Ma old as on crust less than 10 Ma old, even allowing for the smaller area of the younger crust. The older crust should be about three times thicker and thus one-third as vulnerable. The combination of fast drifting, nominally thick lithosphere, and high volcanicity at first appears fatal to the concept of plate vulnerability. However, every criterion indicates that

Distribution of types of oceanic islands on the Pacific plate on crust of different ages. Numbers in parentheses are adjusted as if the youngest period was represented by the same area of crust as the others.

Age (Ma)	Relative area	Active volcanoes	High islands(H)	Banks(B)	Atolls(A) and guyots(G)	BAG	BAGH
0–10	1	2(8)	2(8)	3(12)	0	3(12)	5(20)
10–50	4	1	5	1	31	32	39
50–100	4	7	22	15	103	118	147
100 +	4	0	14	52	462	514	557

midplate volcanism is ordinarily associated with midplate swells and that they are caused by thermal rejuvenation. If so, the lithosphere where the volcanism takes place is about the same thickness regardless of its age.

The graph on this page shows the cumulative number of different types of islands on the Pacific plate as a function of age of the crust. The plot is semi-logarithmic, so a constant rate of accumulation would give a straight line for each type; that is roughly what is seen. In concept, active volcanoes are distributed throughout the plate and they produce islands at a constant rate. The number of volcanoes that are active does not accumulate, in fact, because they rapidly become extinct. The number of extinct high islands likewise does not fill the seas because they rapidly subside. The only types of islands that actually accumulate as the Pacific plate drifts along are the ancient ones: submerged banks, atolls, and guyots. Even the banks would be infrequent if most of them were not recently drowned ancient atolls.

With reasonable assumptions, it is possible to see whether the present rate and distribution of volcanism could yield the observed distribution of ancient islands. Assume that the active volcanoes yield islands in 10^6 yrs

The cumulative number of islands of different types on the Pacific plate as a function of age of crust.

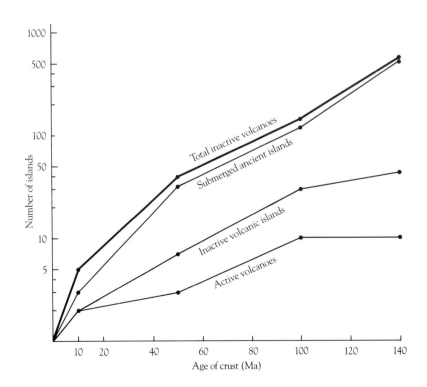

Number of islands calculated for a hypothetical plate having two hot spots each on 10-Ma and 50-Ma crust, and
three each on 70-Ma and 100-Ma crust. Solid triangles
represent active volcanoes from the hot spots; open triangles represent high islands formed by those volcanoes. In
ten million years, the high islands sink and become low
islands and then guyots or atolls. The curve beginning at
each black triangle shows the accumulating number of
ancient islands formed in this way on crust of each age.
The dashed line shows the cumulative number of islands
actually observed on the Pacific plate.

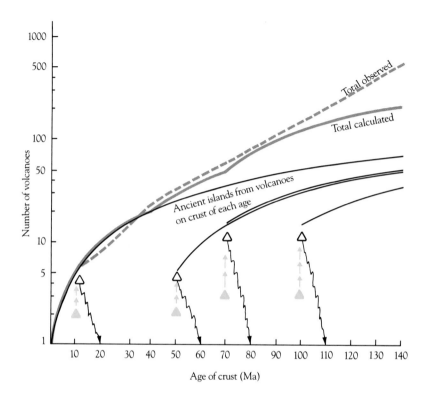

and high islands disappear in 10^7 years—which are merely simplifications
of many observations. Put two hot spots on 10-Ma crust, two at 50 Ma,
three at 70 Ma, and three at 100 Ma—which is a reasonable approximation of observations. Finally, assume that each hot spot produces 2.5 volcanoes in 10 Ma—which is about the ratio between active volcanoes and
high islands. The graph above shows that the model corresponds reasonably well with the observations for the last 80 Ma. The present pattern,
distribution, and rate of volcanism could have produced the ancient islands on most of the Pacific crust.

For older crust, there are far too many ancient islands to have been
produced by volcanism at present rates. The same type of modeling can be
used to determine how intense the volcanism had to be to yield the observed 557 ancient volcanoes instead of the 210 expected from present
volcanism. For example, if all the excess of 347 were produced by some
ancient volcanic spasm on crust less than 10 Ma old the intensity had to
be seventy times what it is now. There would have been 140 active volcanoes on or near the young edge of the Pacific plate. However, many of the

surplus guyots have been sampled and they were not active on young crust but rather on crust from 13 Ma to 64 Ma old. This was about the same crust, now 70 Ma to 110 Ma old, over which the abyssal flood basalts flowed in this region. If we assume that the extra ancient islands were built at a constant rate during this 50-Ma period, the excess number of active volcanoes needed was only eight. The intensity of midplate volcanism, therefore, needed to be about seven times the present rate on crust of that age. Assuming that the present plumes existed then, the excess number could have been produced in two ways, either by greatly multiplying the number of plumes, or by slowing the plate to 1 or 2 cm/yr. The latter phenomenon would have concentrated more volcanoes on the plate without increasing the plume discharge. There is no way to distinguish between the possibilities, but it is noteworthy that the extrusion and intrusion of the abyssal flood basalts in itself required a great increase in the discharge of lava. The excess number of ancient islands merely strengthens the evidence of extraordinary volcanism in the Pacific in Cretaceous time.

4 *Sea Level*

The coastal zone of an island is the locus of a wide range of distinctive phenomena that may be preserved in the geological record. To pick the most obvious examples, coral reefs grow only in shallow water and waves attack only the shoreline. In contrast, the biological and geological zones above and below sea level are much less sharply defined. Thus the indicators of sea level provide the best hope of unraveling the history of vertical movements of islands. Islands are dip-sticks that record the changing distances from the sea floor to the sea surface. Unfortunately, the record locally is the same if the sea floor goes down or the sea surface goes up, and the history of sea level changes is not necessarily informative about causes.

With regard to the vertical movement of islands relative to sea level, three factors are known to be geologically important, namely, cooling of tectonic plates, global fluctuations of sea level, and isostatic compensation for loading and unloading. Probably, local relative sea level is also affected by such factors as mantle convection and changes in rotation of the earth, but, if so, the effects have not been identified for certain in the geological record.

We have already considered the most important factor that influences the long-term vertical motion of islands. They stand on a lithosphere that is sinking at a rate determined by its age or its thermally rejuvenated age. Lithospheric cooling can pull an island down 1000 m in less than 10 Ma and 2500 m in 60 Ma. No other known or suspected effect is as large, so, in the absence of contrary evidence, it can always be assumed that an island is sinking relative to sea level. However, the rate of thermal subsidence decreases rapidly from 36 centimeters per thousand years to a com-

Coral exposed by an unusually low tide on Palau.

mon value of about a tenth of that, and other phenomena cause smaller but much more rapid fluctuations in relative sea level. Consequently, in the short term, an island may emerge or may submerge more rapidly than expected from thermal subsidence. Nonetheless, the effects of thermal subsidence are firmly established, and any deviation from the expected rate is in itself an anomaly that needs explanation. The factor of thermal subsidence, in short, is reasonably well isolated.

The second known factor is global change in sea level. Such a change is called *eustatic,* meaning that the whole sea surface fluctuates by the same amount. The causes and characteristics of eustatic changes will be discussed in this chapter.

The third known factor is isostatic compensation, which acts in response to something else. If thermal subsidence tends to pull an island under water, the additional volume of displaced water will tend to buoy it up—thus reducing the expected change in relative sea level. If a eustatic drop in sea level tends to elevate an island, the lessening of buoyancy will tend to pull the island down—thus also reducing the expected change in relative sea level. Fortunately, the effects of isostatic compensation in many circumstances are qualitatively well known and can even be calculated with confidence. In sum, the most uncertain factor affecting the apparent vertical movement of islands is the eustatic shift of sea level.

RELATIVE CHANGES

The position of sea level varies every second because of waves, tides, and longer-term oscillations. Thus, a useful "level" can be established only by averaging measurements over a year or more; this has been done with tide gauges in hundreds of places for as long as two centuries. Annual mean values are accurate to a millimeter, and with such accuracies and periods of observation it has been shown that many tide gauges have moved relative to a mean sea surface.

For periods longer than a century or two, it is necessary to use "tide gauges" derived from historical or geological rather than physical observations. The discovery of the ruins of a Greek temple in water five meters deep would be historical evidence of a local change in relative sea level. Such evidence is rarely accurate to more than a few meters; for example, the temple has subsided at least 5 m but may have gone down much more if its base was above sea level initially.

Left: This champignon, or "mushroom," on Aldabra in the Seychelles was shaped by the erosion and dissolution of rock at sea level. *Right:* Fossil coral protrudes from a cliff well above sea level on Aldabra.

The important geological indicators of changes in sea level are of two types, biological and erosional-depositional. There are many organisms that live only within a few meters of sea level, but by far the most important for our purposes are reef corals. Elevated coral reefs are found widely on the unstable islands of subduction zones and rarely elsewhere. They are certain indicators of elevation, and they can be dated by isotope geochemistry. Unfortunately, coral reefs are not such accurate physiographic indicators of submergence, because they are usually capable of growing up to fill the gap to sea level. However, as Darwin pointed out, coral reefs are alive and some die when they are submerged. Moreover, drilled reefs give stratigraphic evidence of submergence. Layers of reef rock have been deposited one on another for as much as forty million years.

The other important geological indicators of sea level are wave-cut benches and terraces and the sea cliffs that rise above them. Waves are remarkably effective in eroding even the hardest rock. Growing volcanic islands in the sea are terraced and cliffed between eruptions. In softer sedimentary rock or the sand and mud of the coastal zone, wave erosion is even faster. Thus it may be assumed that any stand of sea level will be recorded by wave erosion within a few meters of sea level. Moreover, waves and currents build beaches, lagoons, and offshore bars, and these may also be preserved as indicators of sea level or, at least, shallow water. These erosional and depositional indicators of sea level are widely observed at tens and hundreds of meters above and below sea level. Thus, even though they are not as accurate as tide gauges, they are accurate enough to be useful.

A problem with coral reefs, wave-cut terraces, and beach deposits is that they are commonly destroyed or buried in a geologically brief time. Oceanic islands themselves do not persist very long, either, so the reefs and benches may endure as long enough to indicate uplift. However, after only a few million years, elevated reefs, for example, may be so modified by solution and erosion as to be unreliable indicators of the amount of uplift.

As it happens, the indicators of sea level are preserved when oceanic islands are submerged rather than uplifted, but not enough atolls and guyots have been drilled to demonstrate a consistent geological history. In contrast, the continental margins have been explored extensively in the search for oil. The geological history of a continental margin, unlike that of an oceanic island, is vastly complicated by deposition of sediment eroded from the adjacent continent. Nonetheless, hundreds of holes have revealed the geological history of the points where they are drilled, and the structure of the wide areas between holes has been tied to them by a technique called *seismic stratigraphy*. Rocks are relatively transparent to low-frequency energy, such as the vibrations produced by explosions, and thus it is possible to map the boundaries between layers of different kinds of rock just as a higher-frequency echo sounder can map the boundary between the sea and its floor. The seismic reflectors can be identified where they intersect the holes.

Exploration has been so intense in many places that relative changes in sea level have been identified. The stratigraphic indicators of these changes consist of patterns of onlap and offlap of coastal and marine sediments, and erosional gaps when there was no deposition. As sea level rises, the shoreline moves inland and beach sands and other coastal sediments lap onto the land. Behind them, marine deposits thicken if they build up to maintain a constant water depth. When sea level falls, the coastal and marine sediments are exposed to erosion, and a temporal gap in the stratigraphic record is created. This simple picture is complicated by the flux of sediment into and out of the region. Nevertheless, the method has been highly successful, although the cause of the fluctuations that it has identified remains controversial.

Local and Regional Warping

Historical and geological indicators of fluctuations in sea level are widespread, and the causes of those fluctuations have been sought for centu-

The Temple of Serapis near Naples.

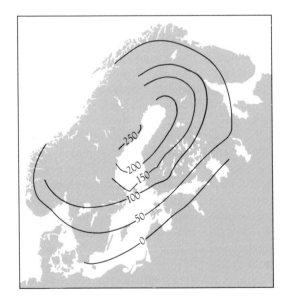

The uplift of Scandinavia (in meters) since the ice cap melted.

ries. The famous Greek temple of Serapis near Naples, for example, was studied by Darwin's mentor, Sir Charles Lyell. Borings by marine molluscs prove that it has been partly submerged more than once. However, the evidence for similar vertical movements is lacking elsewhere in the area, so the cause is local. It is, we now know, the swelling and detumescence of a nearby volcanic region. Likewise, the uplift of Scandinavia was long ago obvious because ancient seaports became unusably shallow, then emerged, and gradually became elevated above a receding shoreline. This uplift extended from Denmark to the northern tip of Norway and from the Atlantic to eastern Finland. Nonetheless, it was a local phenomenon with a local cause. During the ice ages of the past million years, the whole region that now has elevated shorelines was covered by a continental ice cap centered in the northern end of what is now the Gulf of Bothnia. The load of the ice on the continental crust made a dish-shaped depression surrounded by a peripheral bulge. When the ice began to melt, the warped rocks began to resume their original shape. At the shrinking periphery of the ice, the sea cut terraces and left dateable marine fossils. By correlating terraces of the same age, it is possible to map the amount and rate of uplift of the deglaciated region. The center has been uplifted 500 m, and the amount of uplift is progressively less toward the edges of the former ice cap. Moreover, exactly the same evidence of differential, regional uplift can be obtained with tide gauges. Near Copenhagen, the sea floor is rising at 3 cm per century; at Stockholm the rise is 50 cm per century, and at the northern end of the Gulf of Bothnia it is 110 cm per century. All these phenomena are also observed in North America, where there was another ice cap.

The most conspicuous evidence for historical changes in sea level is all associated with local or regional causes, but geologists have tended to seek global causes for more ancient uplifted terraces and reefs. Thus, the abundant evidence of elevated shorelines on Pacific islands has been attributed to several eustatic changes in sea level. In fact, most if not all are caused by the warping of converging tectonic plates or by local loading by young volcanoes (as explained in Chapter 8). The apparent lack of evidence for higher eustatic sea levels in ancient times is a puzzle, because the melting of existing glaciers would produce it, and the glaciers have not always existed. However, when glaciers were much more extensive, sea level was lower and left clear evidence to prove it.

EUSTATIC CHANGES

The many factors that affect sea level are of three types, namely, changes in the density of sea water, changes in its mass, and changes in the shape of the sea basin. As to the first type, tide-gauge records indicate regional changes in sea level that vary with the seasons. In Hawaii, for example, summer heating and expansion of surface waters causes sea level to be 18 cm higher in the early fall than in the spring. Reasonable historical variations in heating, salinity, and air pressure can cause other regional or even hemispheric heating variations as great as 20 cm. Variations due to this cause during geological time may be larger; heating the whole ocean by 10°C would raise sea level by about 6 m, and the sea certainly has been much warmer than it is now. Changes in sea level from variations in density alone are not large compared with other causes, but they occur constantly and it must be assumed that, in addition to the rapid movement of waves and tides, sea level itself is fluctuating hemispherically or globally on a time scale of months or years.

The second cause of change in sea level is variation in the mass of the sea. Changes in density are minor, so changes in mass can be equated to changes in volume. These can occur by the transfer of water between the sea and the interior of the earth, the atmosphere, rivers and lakes, ground-

Distribution of terrestrial water (10^6 km^3)

Mantle	4,600,000
Ocean	1,350
Groundwater	200
Ice (at present)	30
Ice (at glacial maximum)	90
Rivers and lakes	0.5
Atmosphere	0.01

water, and ice. The relative importance of the possible fluctuations is indicated by the distribution of terrestrial water, shown in the table on this page.

No other planet has an ocean, and there is no reason to believe that the earth had one in the early days when, like the other planets and the moon, it was still being bombarded by meteorites. Instead, it is generally accepted that the ocean has evolved gradually by outgassing from the mantle. The number in the table is based on the ordinary assumption that the mantle has the composition of stony meteorites; if so, it clearly is an adequate source for all surface water. What is not agreed upon is the rate at which the ocean grew. Geological evidence indicates that the salinity of the ocean has not varied enough to produce major biological or sedimentary changes. This argues for a relatively constant rate of growth. For present purposes we are concerned only with the last 100 Ma, because there are no older oceanic islands, or even guyots. In that time, growth of the ocean at a constant rate, say, 0.1 cm per thousand years, would have raised sea level by 100 m. It is uncertain that any such rise took place; even if it did, it would have few geological effects. It would not drown the continents because their level is adjusted by isostasy to such slow and persistent changes in sea level. It would not affect the erosion of islands or the life of coral reefs because it is much too gradual. The only effect of any consequences with regard to islands is that ancient guyots might be 100 m deeper than otherwise expected. No such effect is observed. On the contrary, as we shall find, ancient guyots are anomalously shallow. If there has been any significant increase in the volume of the ocean in the past hundred million years, the effect on sea level has been completely masked by other effects.

Turning to other sources and sinks of seawater, it is evident that the atmosphere, lakes, and rivers do not, and moreover cannot, store enough water to have any important effect on sea level. There is an extremely rapid flux of water to and from the ocean through the air and rivers but the volume is insignificant. The relation is just the opposite for groundwater buried in ancient sedimentary rocks. That volume is significant but the flux, caused by erosion and deposition, is extremely slow, so probably it, too has not affected sea level in the geological period of interest.

That leaves ice. If the Antarctic and Greenland ice caps were to melt, sea level would rise about 50 m even after isostatic adjustment. (That is one of the reasons to be concerned about global heating through excessive burning of fossil fuels and the greenhouse effect.) The continental glaciers repeatedly were much more extensive in Pleistocene time than now, and it is estimated that sea level was lowered 150 m to 200 m several times. The rates and geological consequences of these extremely important fluctuations in sea level will shortly be discussed. Suffice it to say for the moment that fluctuations at such rates and on such a scale have not been typical of most of geological time.

The third cause of change in sea level is variation in the shape of the ocean basins. For example, the growth of submarine volcanoes and deposition of marine sediment elevate sea level. Geologists used to assume that these effects were cumulative and, indeed, it was initially proposed that guyots are deep because eons of slow sedimentation had gradually raised sea level. However, the discovery of plate tectonics eliminated that assumption. The sea floor is relatively young, and all the older marine volcanoes and sediment have been swept onto the continents or down into the mantle. It is now reasonable to assume that for each lava flow or grain of sand that enters the ocean, one is lost by subduction in trenches, so sea level is unaffected by these processes.

A global balance of evaporation and precipitation may also be assumed to leave sea level unaffected, but occasionally there may be an imbalance because of a change in the shape of the ocean basins. This occurred about 5.5 Ma ago, when the Mediterranean region was tectonically isolated from the Atlantic. The global average rate of evaporation is about a meter per year and, in the Mediterranean basin, that was not balanced by rain and rivers. In a geological trice, much of the Mediterranean became a salt-encrusted desert and the rivers were cutting enormous canyons down what had been (and would again be) the continental slope. The evaporated water of the Mediterranean Sea had to be transferred to the world ocean and thereby raise sea level, but, at most, the rise was only 10 m. That rise disappeared in a geological instant when the Atlantic spilled over the Straits of Gibraltar in a gigantic waterfall. Isolation of ocean basins appears to be rare, but it can cause rapid eustatic changes in sea level.

The only other known way to alter the volume of an ocean basin significantly is by a change in the volume of midocean ridges. This may occur by a change in the length of spreading centers or in the rate of spreading. Ridge crests are constantly being shortened by subduction and lengthened by propagation. Inasmuch as the crests are shallow and displace water, these variations affect sea level. Unfortunately, the effects are difficult or impossible to quantify because subducted ridges cannot even be counted let alone measured.

In contrast, the effects of a change in spreading rate can be calculated with confidence. For the purposes of illustration, consider the consequences if the spreading rate became infinite. The ocean basins would all be about 2500 m deep and the continents would be deeply flooded. Clearly, a more probable increase in spreading rate would also flood the continental margins and, from the relation of sea-floor depth to age, the amount of flooding could be calculated. As it happens, the continental margins about 100 Ma ago were flooded, and the cause is believed to be accelerated sea-floor spreading.

The Past 35,000 Years

Fluctuations in sea level during the past 35,000 years are relatively easy to determine, because fossil shorelines are abundant and they can be dated by the radiocarbon content of shells and wood. The fluctuations differed from place to place because of noneustatic effects. However, global eustatic effects are clear. Sea level from 35,000 to 30,000 years ago was near the present level. By 16,000 years ago, it was 130 m lower because of the growth of glaciers. About 14,000 years ago, sea level began to rise rapidly, but it slowed 7000 years ago and has been relatively constant for the last 5000 years. Reginald Daly, who first realized the effects of glacial-eustatic fluctuations on islands, thought sea level had been about 5 m higher than now in relatively recent time. There is a narrow bench and a nip in sea cliffs at that height on some islands, but it is not now thought to be due to eustatic change.

The maximum rate of eustatic change was about 1000 cm per thousand years for periods of 5000 to 10,000 years. Except possibly for the

Sea level for the past 35,000 years. The solid curve is the depth below present sea level of the former Atlantic coast of the United States. The dashed curve is for the coast of Texas.

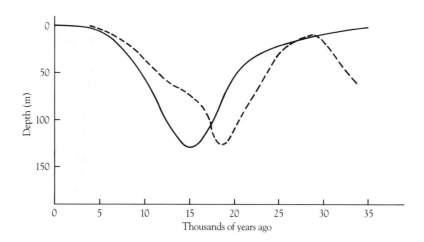

A wave-cut bench and nip in a sea cliff in the Gambier Islands. The bench is about one meter above the present mean sea level.

evaporation and flooding of the Mediterranean basin, these glacial-eustatic fluctuations are by far the fastest known. They are roughly 1000 times faster than fluctuations caused by variations in the speed of sea-floor spreading. They are 30 to 50 times faster than the thermal subsidence of the sea floor. Thus it can be assumed that, whenever there are large continental ice caps, glacial-eustatic fluctuations will wholly dominate the movement of sea level relative to an island.

The principal effects of rapid fluctuations of sea level on islands depend on the efficiency of wave erosion and the life of coral reefs. With regard to waves, the entire depth of the ocean is affected by wave motions such as tsunamis and tides. However the abyssal motion is capable of no more than stirring sediment. Almost all wave energy of geological interest is expended by the breaking of surface waves at the shoreline in depths of less than 10 m. During a long still-stand of sea level, waves should cut a gently sloping rocky terrace on which wave energy is lost to sediment transport and friction. Thus, the most important effect of rapid fluctuations in sea level is to expose a coastal band to the shallow depths of vigorous abrasion by waves. Most continental and insular shelves are less than 130 m deep, and consequently the shoreline has swept across them twice in the last 35,000 years.

Coral reefs can grow so rapidly that they should always be as high as the highest stand of sea level. If there had been a eustatic sea level 5 m

In New Guinea, the repeated eustatic rise and fall of sea level combined with the steady uplift of the shoreline to create a giant staircase of dead reefs.

The Past Million Years

Continental ice caps have developed and vanished several times in the history of the earth, so their existence at present is not proof of secular cooling. Nonetheless, a trend toward the growth of ice caps has continued now for 30 million years. Some ice has existed in Antarctica during that entire period, and the eastern ice sheet developed by 16 Ma ago. The western Antarctic ice sheet followed 5.5 Ma ago. In the Northern Hemisphere, glaciers began about 2.4 Ma ago, and during the last million years or so continental ice caps formed repeatedly. It appears that the fluctuations are strongly influenced by periodic variations in the earth's tilt and precession, which, in turn, influence solar heating at the earth's surface. Inasmuch as the astronomical factors are well known, it appears possible to date glacial fluctuations when other methods are lacking.

higher than at present, all atolls should now rise that high, but they do not. On the other hand, all coral islands were 130 m high 16,000 years ago. Many atolls that are presently drowned also were then exposed, to lesser heights. Just what then happened to the reefs is conjectural and will be discussed after consideration of repeated glacial-eustatic fluctuations in Pleistocene time.

The influence of astronomical cycles on eustatic changes in sea level has been confirmed by dating elevated coral reefs on Barbados and New Guinea. Both these islands have been uplifted gradually, as a consequence of the horizontal compression in subduction zones, and the reefs resemble giant staircases. Both the tops and faces of the individual reefs in each sequence have been dated and found to correspond to the schedule of astronomical fluctuations of insolation. It is necessary to make only the simplifying assumption that the uplift has been at a constant rate in order to generate a complete history of sea level for the last 140,000 years. This type of study indicates that sea level was a few meters higher than now about 125,000 years ago but has since been lower until recently. In the interval, it fluctuated up and down by 40 m to 60 m six times. Every time sea level went down, a reef died and was gradually elevated to become the bottom step in the staircase. Meanwhile, when sea level rose again rapidly compared with the elevation of the land, a new reef grew upward and in turn died to become a new bottom step.

Glaciers leave obvious evidence of their existence, but the record on land is complicated by the fact that younger glaciers may override the traces of older ones. From this continental record, geologists concluded that there were four glacial and interglacial periods in Pleistocene time, about the past million years. However the astronomically determined periods of low insolation were far more frequent, and the waxing and waning of glaciers probably was correspondingly variable. For present purposes, it may be assumed that sea level fluctuated by significant amounts scores of times. Likewise, reefs and atolls were repeatedly exposed and elevated by more than 100 m and perhaps as much as 200 m.

What happened to the atolls when they were elevated? The answer is still uncertain, perhaps because there were large local variations. At present, it is apparent that some uplifted coral reefs are being dissolved at or just above sea level. Thus, the uplifted atolls of Pleistocene time might have been dissolved away and replaced by a new reef whenever sea level again rose. This hypothesis is supported by the correlation between rainfall and the depths of atoll lagoons. Rain can slowly dissolve elevated reefs, and the depth to which an elevated atoll was dissolved thus might vary with rainfall. The broad, relatively flat floor of the central lagoon of an atoll presumably rests on or above the level of the surface of solution when the atoll was elevated. Atoll lagoons range in depth from 0 m to 150 m, and tropical rainfall in the open ocean ranges from 40 cm to 520 cm per year. Thus, there is great variability, and it is striking that lagoon depth within many island groups varies systematically with rainfall. However, there is no systematic variation from one group to another. The deepest lagoon in the Ellis Islands, for example, is about the same depth as the shallowest one in the Maldives although the Ellis group has the greater rainfall. Thus, factors other than rainfall must be very important in the history of elevated atolls.

The best kind of information available about the effects of swinging sea level comes from drilling, but it is regrettably sparse. The Pleistocene reefs of three atolls, Eniwetak, Bikini, and Mururoa, have been drilled in connection with nuclear tests. The stratigraphy may be summarized as follows:

0 m–10 m   coral less than 6000 years old
Erosional unconformity
10 m–20 m   coral 120,000 to 150,000 years old
Erosional unconformity
20 m–50 m   coral more than 500,000 years old
Erosional unconformity
50 m–82 m   undated coral
Erosional unconformity

The reefs apparently had a consistent history, although data are scant and generalizations may be risky. During the past 6000 years, 6 m to 10 m of coral has grown upward during a period of relatively stable sea level. Between 120,000 and 6000 years ago, when sea level was lower than now,

Left: An eroded elevated reef at Kaukura Atoll, in the Tuamotus. *Right:* The channels in the lagoon floor of Mataiva Atoll are evidence of erosion during a former period of relative emergence.

although fluctuating, the reef was eroded. Between 150,000 and 120,000 years ago, when sea level was high, from 10 m to 12 m of coral built up. Before that was a period of erosion beginning more than 500,000 years ago. The earlier events are obscure, but there were two more cycles of alternating erosion and deposition, giving a total of four. The history indicates several facts about the relationship between these atolls and sea level. First, coral accumulated only during high stands of sea level. Second, intervals of erosion were much longer than those of accumulation. Third, coral 120,000 years old is at only about 10 m and thus it was not dissolved very deeply, if at all, when sea level was at its lowest. Nor, during the last 500,000 years, has solution extended below 20 m.

The sparse but high-quality data from drilling suggest that not very much happened to elevated atolls. Presumably the same minor amount of solution has taken place on existing elevated atolls, but no major erosion. The drilling data also suggest a rise in sea level relative to the atolls. In New Guinea, the gradual elevation of the sloping land and fluctuation of the sea produce one reef beside another. In the main ocean basins, the gradual submergence of the sea floor and fluctuation of the sea produce one reef on top of another, with periods of erosion in between.

Tertiary and Cretaceous Sea Level

The Tertiary Period, extending from about 65 Ma to 1 Ma ago, had important fluctuations in sea level, according to the interpretation of the seismic stratigraphy of continental margins by Peter Vail and colleagues. They believe that sea level was generally higher than now for the first half of Tertiary time and that the most important single event was a rapid drop in Oligocene time, about 30 Ma ago. The onlaps and offlaps of the stratigraphic record may indicate both changes in sea level and changes in rate of subsidence of continental margins. In any event, the actual amount of these relative changes is conjectural. Nonetheless, it appears probable that fluctuations in sea level during Tertiary time occasionally exposed atolls and the insular shelves of volcanic islands to erosion. These do not seem like favorable conditions for the growth of atolls, but, curiously enough, most if not all existing atolls are of Tertiary age.

The geological history of sea level in Cretaceous time, from 135 Ma to 65 Ma ago, is notable for a rise that caused broad shallow seas to spread even to the interiors of continents. Seismic stratigraphy indicates few fluctuations in this gradual rise. Consequently the islands of Cretaceous time were rarely elevated and exposed to accelerated erosion. Instead, they were ordinarily submerged rather rapidly by the combination of thermal subsidence and a rise in sea level. These circumstances presumably would have been highly favorable for reef organisms of Tertiary time, but they were not for Cretaceous reef builders. Certainly the Cretaceous reefs did not ordinarily, if ever, grow up to become the platforms of Tertiary atolls. Instead they died and went down with the numerous Cretaceous islands that became the guyots of the western Pacific.

5

Growth of Isolated
Volcanic Islands

The composition of igneous rocks affects their physical properties—notably, their viscosity—and thus largely determines the morphology of subaerial volcanoes. A broad, gently-sloping, shield volcano is produced by fluid flows of basalt, whereas a small, steep-sided cinder cone is normally produced by viscous, explosive andesite or rhyolite. In the ocean basins, almost all the rocks are basalts, so, although the exceptions are interesting, there is little variation in composition to affect the growth of submarine volcanoes. On the other hand, analyses of the rocks provide almost everything known about their source and history before eruption or intrusion.

It is a gross simplification, but adequate for our purposes, to say that ocean basins are made of two types of volcanic rock—*tholeiitic* and *alkalic basalts*. Tholeiite is lower in the alkaline elements potassium and sodium. It is a common rock on some islands, notably Hawaii, where it is voluminous. The basalt that forms the top layer of the oceanic crust by filling in spreading centers is a distinctive tholeiite that is similar in some ways to basaltic achondritic meteorites. Oceanic tholeiites are the most common volcanic rocks on earth, but they were discovered only twenty years ago, by A. E. J. Engel. They are now the subject of so many studies that petrologists call them by the acronym MORB, meaning *midocean ridge basalts*.

Alkalic basalts are characteristic of large seamounts as well as islands. They are relatively enriched in the same elements and isotopes that are concentrated in continental crust compared with the interstitially more primitive rocks of the mantle.

The physical properties of the ocean and atmosphere also affect the growth of volcanoes. To lava at 1200°C, the temperature range in the sea and air is relatively unimportant—one place is just like another. Pressure is another matter. The pressure on the deep sea floor is about 500 atmospheres, which is enough to confine most of the dissolved gas in magma. The same magma, erupting on an island, may release so much gas that expanding bubbles make up much of the volume of the lava, and the rock that solidifies is pumice. Higher concentrations of dissolved gas may cause the bubbles to coalesce and explode the lava into volcanic ash. Somewhere between the deep sea and the air, tiny bubbles must begin to form in lavas in the conduits of volcanoes. If the top of the growing volcano is still underwater, the lavas solidify with small bubbles inside (unless they explode). The presence of such bubbles greatly affects physical properties of volcanoes—particularly, their density—that may influence their growth.

Another important factor is isostasy. As the mass of a volcano grows, it sinks until it is isostatically supported in some way. Just what happens depends greatly on whether support is regional or local, but, in any event, a volcano must extrude enough material to build a root if it is to become an island.

FINDING SUBMERGED VOLCANOES

For a long time, volcanologists were skeptical of sailors' reports of volcanic eruptions in deep water, despite eyewitness accounts of steam, discolored water, and pumice and ash floating at the surface. The known properties of steam and carbon dioxide, the common gases in lavas, seemed to rule out explosions at such pressures. One oceanographer, Sir John Murray, who had a world of experience with the quality of selected marine observations, was more receptive. Indeed, he remarked that reports of an eruption in the equatorial Atlantic in 1852 were good news, because there was no island nearby and Britain needed another coaling station. However, the eruption, if such it was, was in 5300 m of water, and the island has not yet appeared.

One reason for skepticism was that the reports were few and scattered, but they could hardly be otherwise when there were few ships and those were mostly confined to trading lanes. Even now, it is not a coincidence that reports from ships of eruptions in deep water have been more numer-

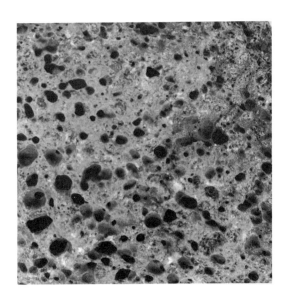

ous in the heavily-traveled North Atlantic than elsewhere. Nor is it chance that most reports from ships were in the nineteenth century, when a lookout posted in the high crow's nest of a sailing ship could see every-thing in a large circle, and a captain had some discretion or curiosity to investigate marine phenomena. Surface indications of eruptions now are observed by pilots who are six kilometers up and can scan enormous areas. One pilot spotted the usual signs of an abyssal eruption in 1955 along the Hawaiian trend far to the northwest of the oldest high island, Kauai, which itself has not been active for millions of years. Excellent charts show that the water depth was 4000 m—provided the pilot knew where he was in the air. Perhaps the clinching demonstration of an abyssal ex-plosion did not come until 1977, when New Zealand Air Force personnel observed turbulent, discolored water while flying in the Tonga-Kermadec region over water 4000 m deep. They dropped a sonobuoy, which de-tected explosions.

In recent decades, scientists have developed methods of detecting active volcanoes in the ocean without being on the site, and four volca-noes have already been found in this brief period. The first method de-pends on the existence of a layer in the ocean at about 800 m where the sound velocity is a minimum. Sound that enters this layer, called the SOFAR channel, travels great distances in it instead of spreading through the whole volume of the ocean. The U.S. Navy has established a network of hydrophones to monitor the channel for explosions. Downed aviators have small SOFAR bombs, which are triggered by pressure at the right depth to be in the channel. Triangulation then gives the location of the aviator.

Oceanic volcanoes erupt explosively in shallow water, and that en-ergy enters the SOFAR channel. However, the U.S. Navy is more inter-ested in the northern than the southern hemisphere, so the SOFAR sys-tem does not give accurate positions in the South Pacific. Nonetheless, in 1967, Rockne Johnson thought he could pinpoint the location of a series of explosions detected by SOFAR southeast of the Austral Islands. In due course, he went there in his own sailboat and discovered an active subma-rine volcano, which he named MacDonald Seamount in honor of the Hawaiian volcanologist. It is only about 100 m deep and may emerge at any time—although not to become a coaling station.

The other scientific method is by seismology, and it does not depend on the occurrence of explosions. As magma rises in a volcanic conduit, it generates small earthquakes, which are sometimes so frequent as to seem a single event, a "harmonic tremor" that lasts for hours. The quakes are below the detection level for distant stations, but French scientists have installed an extensive network of stations in Polynesia. With this system,

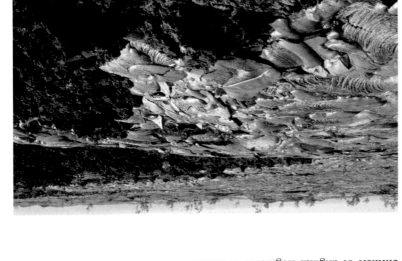

Pahoehoe and aa.

three active submarine volcanoes have been discovered southeast of Tahiti.

Perhaps other active volcanoes may be expected, because the plate
tectonics model requires a hot spot updrift from each line of islands. Mac-
Donald Seamount presumably marks the hot spot for the Austral Islands;
the cluster of volcanoes southeast of Tahiti is over the Society Islands hot
spot; and so on. The hot spots that generated other lines of volcanoes,
such as the Marquesas and Pitcairn lines, have yet to be found. Either they
are not active at the moment or they are still undetected.

LAVA FLOWS

Submarine eruptions have been observed much less than those on land,
which are quite well known. The basalt flows on Hawaii occur in two
modes that are distinctive enough to have Polynesian names, *pahoehoe* and
aa. Volcanologists have found that the mode has nothing to do with the
composition of the lavas; pahoehoe can change into aa, although not the
reverse. Pahoehoe has smooth ropy or billowy surfaces, not unlike poured
chocolate fudge; it is the hot, fluid, volatile-laden lava that ordinarily
pours out of Hawaiian volcanoes. When pahoehoe becomes more viscous
by cooling or degassing, it changes to aa, which resembles a field of
clinkers or angular fragments of rocks.

Left: Lava flowing in shallow water near Hawaii.
Right: Cooled pillow lava.

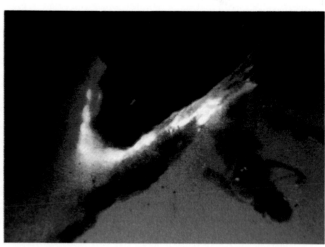

Pahoehoe is very fluid. It flows into and fills valleys and any slight depression, with the result that Hawaiian volcanoes seem very smooth from a distance. A flow quickly develops a crust on all sides to make a tunnel, through which it moves as the tunnel continues to thicken and extend. The main feeder tunnel, or lava tube, branches at the end of a flow just like the distributaries in a river delta. The toes that extend from the small tubes overlap each other to fill transient low spots, and in this respect the end of a flow rapidly simulates the geological history of a great delta. Pahoehoe has flowed more than 30 km twice on Hawaii in histori-cal times. Aa flows have a central core of dense, pasty liquid that is thermally insulated by a layer of jagged, spiny, sharp-edged pieces of clinker, so aa also flows in a tunnel of sorts, but not such distances as pahoehoe.

Submarine lava flows also take two forms, and they are not too dissim-ilar from pahoehoe and aa after due allowances for the different properties of air and water. The equivalent of pahoehoe is called *pillow lava* and it forms only underwater, although not necessarily seawater. Some of the best exposed pillow lavas are in Iceland, where they formed in lakes melted within overlying Pleistocene glaciers, which have since vanished. Pillow lava has been observed to form where pahoehoe flows enter the ocean. In 1905, the British geologist Sir Tempest Anderson was on a ship off the Samoan island of Savaii watching the eruption of the volcano Matavanu. He used a ship's boat to move in toward shore for a close look at a stream of lava flowing slowly and passively into the Pacific. The water was clear and remained calm, so the boat moved over the submarine extension of the flow. Protrusions, or pipes, of lava grew out from the front of the flow and shortly swelled into the typical sacklike forms of pillow lava. The water chilled the lava so rapidly that, when hot lava broke through a crust at some point, it was enveloped in its own crust, which expanded like a loaf of bread until it too was broached. Sir Tempest was a dedicated scientist, but, as he later reported in London, he reluctantly concluded his observations when the pitch used for caulking the boat began to melt. As it happens, his problem arose because he was directly over the flow. More than half a century later, scuba-diving geologists watched and shot a movie of pillows forming off Hawaii. They were close but to the side.

Hyaloclastite is the nearest submarine equivalent to aa. It is a fragmental material that was first identified in Iceland, where it had formed in profusion by explosive granulation under the Pleistocene glaciers. Thick beds of similar hyaloclastite have been found by deep-sea drilling, and the material is commonly dredged on seamounts and cored in abyssal sediments near them. It has not been seen to flow underwater, but what are obviously flows of a sort have been found on the peaks of small seamounts off Mexico. In 1980, Peter Lonsdale and Rodney Batiza used the research submersible *Turtle* to dive on very young seamounts rising 800 m to 1200 m above the sea-bottom depth of 3000 m. They found flows of pillow lava and sheet flows, which appear to be a more fluid equivalent of the bulbous pillow lava. In addition, they found hyaloclastite flows or stone streams leading downslope from pillow lava, but also capping a seamount without associated pillow lava. It appears that, as with pahoehoe and aa, the mode of flow is controlled by viscosity, and the critical viscosity may be exceeded either in flows or while the lava is still in a vent. Aa sometimes flows directly from volcanoes in Hawaii when magma has been stirred or "gargled" and degassed before it flows out. Lonsdale and Batiza conclude that the hyaloclastite flows are facilitated by explosive vaporization of sea water. If so, the phenomenon could occur only where the pressure was less than the critical pressure of water, or not much deeper than 2000 m.

The Density of Volcanic Rocks

Studies of rocks dredged off the Reykjanes Ridge and the eastern slopes of Hawaii show systematic variations in vesicularity (bubble content) and density that are closely related to water pressure. The two studies were made by dredging fresh volcanic bedrock that had not fallen or flowed from some shallower depth. In short, the rocks had not come from the peak of a volcanic cone but from rifts that were erupting at the depth where sampling occurred. Appropriate sampling was relatively easy on the

Cape Verde Islands　　　　　　　Dakar

0
1000
2000 fathoms
— 3500 fathoms
2000　　　　　　　　　2500　　　　　　　　3000

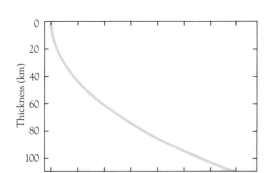

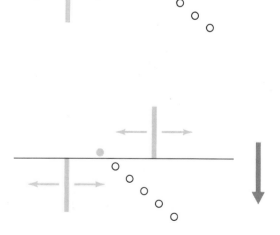

higher than at present, all atolls should now rise that high, but they do not. On the other hand, all atolls were coral islands 130 m high 16,000 years ago. Many atolls that are presently drowned also were then exposed, to lesser heights. Just what then happened to the reefs is conjectural and will be discussed after consideration of repeated glacial-eustatic fluctuations in Pleistocene time.

In New Guinea, the repeated eustatic rise and fall of sea level combined with the steady uplift of the shoreline to create a giant staircase of dead reefs.

lagoon in the Ellis Islands, for example, is about the same depth as the shallowest one in the Maldives although the Ellis group has the greater rainfall. Thus, factors other than rainfall must be very important in the history of elevated atolls.

The best kind of information available about the effects of swinging sea level on atolls comes from drilling, but it is regrettably sparse. The Pleistocene reefs of three atolls, Eniwetak, Bikini, and Mururoa, have been drilled in connection with nuclear tests. The stratigraphy may be summarized as follows:

> 0 m–10 m   coral less than 6000 years old
> Erosional unconformity
> 10 m–20 m   coral 120,000 to 150,000 years old
> Erosional unconformity
> 20 m–50 m   coral more than 500,000 years old
> Erosional unconformity
> 50 m–82 m   undated coral
> Erosional unconformity

The reefs apparently had a consistent history, although data are scant and generalizations may be risky. During the past 6000 years, 6 m to 10 m of coral has grown upward during a period of relatively stable sea level. Between 120,000 and 6000 years ago, when sea level was lower than now,

The new island of Surtsey rises from the sea near Iceland.

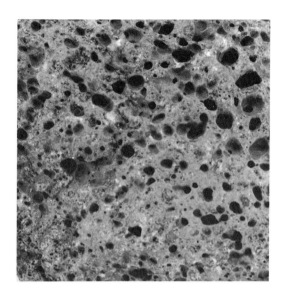

Reykjanes Ridge, just south of Iceland, because it is a spreading midocean ridge and it deepens slowly toward the south. Sampling was more difficult off Hawaii, where flows and slides move fresh rock down steep slopes. It was achieved by sampling only fresh rock along the submarine extensions of rift zones that had been mapped on the island.

The Reykjanes study shows that volume percent of vesicles increases from 5 percent at 1000 m to between 40 percent and 50 percent at 100 m; the density of the rocks correspondingly drops from 2.8 gm/cm^3 to about 2.0 gm/cm^3. These values apply to a midocean-ridge tholeiite. The Hawaiian rocks are also tholeiites, but of the type characteristic of islands. The Hawaiian samples extend down to 5000 m, where their density is about 3.0 gm/cm^3 and the volume percent of vesicles is zero. The values are the same at 4000 m, although a few vesicles 0.1 mm in diameter are visible. Vesicles occupy up to 5 percent of rock volume at 2000 m, but the density still exceeds 2.9 gm/cm^3. At 1000 m, vesicles occupy 10 percent and density is about 2.8 gm/cm^3. At shallower depths, as the pressure rapidly decreases, vesicles increase in size to 0.7 mm diameter, and their volume increases beyond 15 percent; at 500 m, the density is down to 2.2 gm/cm^3. There is no way to relate these results to abyssal explosions, which are likely whenever incandescent rock contacts water. Nevertheless, the Hawaiian rocks are the normal kinds for oceanic islands, so it seems reasonable to assume that they are representative products of oceanic volcanism.

BIRTH AND GROWTH OF A VOLCANO

We can now visualize the initiation of volcanism on the deep sea floor. A crack opens or a caldera sags and lava in pillows and sheets begins to flow outward. If the crust is shallow, hyaloclastites form at the ends of the fluid flows, but they are not abundant. Judging by the growth of small volcanoes on land, a sizeable cone may be built in this way within a few years, but only if the new volcano is growing on young, sediment-free crust. If volcanism begins on old crust, the circumstances are different. The rising magma, with a density of about 3.0 gm/cm^3, reaches a boundary between the original volcanic surface of the sea floor and the soft, wet, low-density sediment above it. It should then flow laterally as a sill under the sediment. Thus, a volcano would appear only after successive intrusions had engulfed enough sediment to form a base.

The shapes of tiny seamounts have been revealed by sonar arrays that form narrow beams whose echoes are reduced to contour maps by computers. They have also been studied with very long-range side-scanning

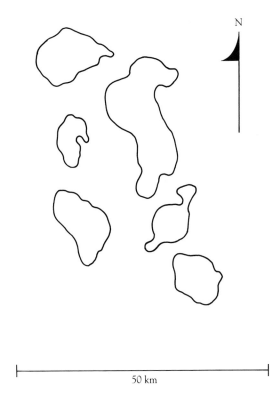

N

50 km

The typical shapes of small conical submarine features. These are in the Philippine Basin.

sonars and research submersibles. Their morphology is highly variable, ranging from gently sloping domes to peaks with slopes of 30°. Some are circular, some arcuate, some complex in plan. One even looks like an H. Most have summit craters but not all, and some have flat tops as though with filled craters. On these tiny volcanoes, 500 m to 1000 m high and 4 km to 6 km in diameter, may be even smaller parasitic cones. Their shapes apparently result from the complex interaction of such factors as conduit geometry (thus the H), magma ascent rates, eruption rates, tectonic activity, crater formation and filling, and the duration of volcanism. The composition of the rocks in the seamounts is not as variable as the morphology, but neither is it constant. Most of the small seamounts that have been sampled are near ridge crests, where they are certain to be young. Batiza has found that many have the same composition as normal midocean-ridge basalt (MORB) and visualizes three ways how this could happen. The volcanoes could merely tap a magma chamber under the spreading center. This seems unlikely, because passage through the long conduit would enhance cooling and crystallization, which would alter the composition of the source magma. A second possibility is that a magma reservoir is encapsulated in the crust, and a volcano grows above it as they drift away together from a ridge crest. The same problem seems to rule this out—the limited source would produce quite different magmas as it was drained. A third possibility presents no apparent problems; The volcanoes may tap the same mantle source as the spreading center but vertically, through the hot interior, instead of laterally, through the cool crust. In any case, it is easy to understand why the tiny volcanoes mostly develop on thin young crust, which is easy for a small volume of magma to penetrate.

Continuing Growth

The history of some old land volanoes can be determined because their interiors are exposed by deep erosion. However, oceanic volcanoes are not eroded in the same way. (See Chapter 6). Thus, at present, the only way to reconstruct the growth of an individual oceanic volcano is to study volcanoes of different sizes and assume that they are equivalent to an age sequence. In essence, different sizes of volcanoes are assumed to be stacked one on top of another. The information we have on intermediate-sized seamounts comes largely from dredged rocks and bathymetric surveys, and it is highly consistent. Hundreds of dredge hauls from seamounts with a relief of 3 km to 4 km indicate that most are constructed of alkalic basalts. The only consistent exceptions are seamounts that certainly or probably were at plate boundaries. For example, Cobb Bank is near the crest of the Juan de Fuca (midocean) Ridge; a seamount on the Cocos (aseismic)

Ridge, like the entire ridge, may have been active when over the Galapagos spreading center; and the Mendocino Ridge was active over a transform fault. These exceptions are composed of tholeiitic basalts that are intermediate in character between MORB and alkalic basalts.

The morphology of the intermediate-sized seamounts is also broadly consistent. Some have very odd shapes, but the majority by far are reasonable approximations of cones. If their relief is only 2500 m to 3000 m and the peaks are deeper than 1000 m to 1500 m, the cones have slopes of 10° that meet the relatively flat abyssal sea floor with little or no transition. However, if the relief is greater, a transition zone with a slope averaging 4° appears. As the volcano grows upward, the height of the upper steep slope is little altered, but the lower gentle slope increases in radius to accommodate the upward growth. It is as though the active volcano rises at a constant height above a growing pile of its own effluvia. This continues until the volcano approaches sea level, where a new regime begins.

EFFECTS OF LOADING BY AN ISOLATED VOLCANO

A small volcano growing on the deep sea floor exerts a load on the lithosphere, but it produces no more perceptible deformation than the load of a person standing on a great bridge. As growth continues, however, the load gradually increases and something has to balance it. What happens depends on the stiffness, of *flexural rigidity,* of the lithosphere, and the amount of the load.

The flexural rigidity of a sheet or plate of material varies with the thickness. This is known from common observations. If a sheet of cardboard partly extends over the edge of a table, it bends down under its own weight; a thin sheet of plywood bends less; a thick sheet bends hardly at all, if it projects only ten or twenty centimeters. Another effect of flexural rigidity is equally common but not quite as obvious. Consider the following experiment. One end of a thin sheet of plywood is bolted to a heavy table, with a third of the sheet extending beyond the edge of the table. The unsupported third will bend downward, and the supported part of the sheet that is nearest the table edge will flex upward. If the same experiment is repeated with sheets of cardboard and plywood of different thicknesses, the same combination of downward and upward flexing occurs. However, the greater the thickness and thus the flexural rigidity of the sheet, the greater the distance that the upward flex extends inward from the edge of the table. With regard to the oceanic lithosphere, the thickness and flexural rigidity increase with age. Thus, the width of the area

flexed by loading increases with the age (or thermally rejuvenated age) of the lithosphere at the time of loading.

Consider now what is observed if the series of experiments is repeated with a modest weight attached to the unsupported ends of the sheets. The width of upward flexing is unchanged for each sheet, but the free edge is pulled lower and the upward flex is correspondingly higher.

Finally, if each experiment were continued indefinitely, the flexing would not change. To the extent that the materials are elastic and their strength is not exceeded, the sheets stay flexed as long as the load remains.

These laboratory experiments concern linear loading and linear deformation. At sea, linear deformation is found beside volcanic ridges or lines of overlapping volcanoes such as the Hawaiian islands. Isolated volcanoes are more like point loads. Such loads flex a central circular area downward and a surrounding larger annulus upward. They differ from the line loads of the experiment only because of the geometry—the width of the area flexed still depends on flexural rigidity, and the height of the upward flexing still increases with loading. However, the annular upward flex is not elevated as much as it would be by a linear load, where the upward flexing may be considered to be the overlapping effects of many closely spaced point loads.

The Discovery of Regional Compensation

The fact that islands can be supported by the flexing of wide areas of crust around them, rather than simply by floating on the denser rock below, was first shown by F. A. Vening Meinesz. In the 1920s, he began decades of gravity measurements in submarines with a compound pendulum of his own devising. (A submarine could dive and stabilize itself in relatively motionless water below surface wave action.) Over the years he found that most of the sea floor was isostatically compensated (supported) locally, like the continents. There were two exceptions. Deep-sea trenches were grossly out of isostatic equilibrium, and Vening Meinesz deduced that powerful horizontal forces were at work in those places. The other exception was the Hawaiian region. The islands were not locally compensated. Instead they were surrounded by a narrow belt whose gravity was less than normal, and it, in turn, was surrounded by a belt where gravity was higher than normal. Vening Meinesz reasoned that the great volcanic islands had been constructed on a rigid plate of lithosphere, and that their load was distributed by the plate to the surrounding region, which was accordingly flexed. He called this *regional* as opposed to local isostatic compensation.

Regional compensation for the load of volcanoes has since been confirmed by gravity measurements for many other island groups and individual isolated islands. Three discoveries particularly stand out since Vening

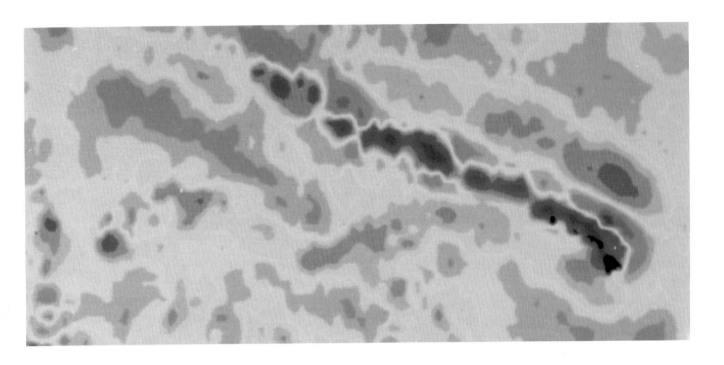

The north central Pacific as shown by satellite-measured gravity anomalies. The great lineations in the right half of the picture are caused by the load (red) of the Hawaiian volcanoes and the parallel flexing downward (blue) and upward (red) of the lithosphere. The overall speckled pattern is caused by oceanic volcanoes, including atolls and guyots.

Meinesz's time. A. B. Watts and colleagues at Lamont-Doherty Geological Observatory showed by three-dimensional analysis that the point load of Great Meteor Seamount, a large guyot in the Atlantic, caused only a quarter of the gravity anomaly, and thus the flexing, that would result from a line load. Second, they showed that the elastic thickness of the oceanic lithosphere (the thickness that influences the flexural rigidity of a plate) increases with age, from negligible near ridge crests to between 20 km and 40 km at age 100 Ma. The material below is also part of the plate, but it is plastic in response to long-term vertical stresses, so local isostatic compensation is associated with a thin elastic thickness. The third discovery was the ability to map unknown seamounts in remote regions of the ocean basins by means of artificial satellites that measure gravity or the elevation of the sea surface to within a few centimeters. Moreover, the relation between lithospheric age and compensation makes it possible to date a seamount without a rock sample. If a seamount grew on young crust, it is locally compensated; so if a known seamount is so compensated, it is the same age as the crust. This last discovery is only beginning to be investigated.

Enough spot soundings had already been made around the Hawaiian Islands for Vening Meinesz to identify a moat as well as an encircling arch.

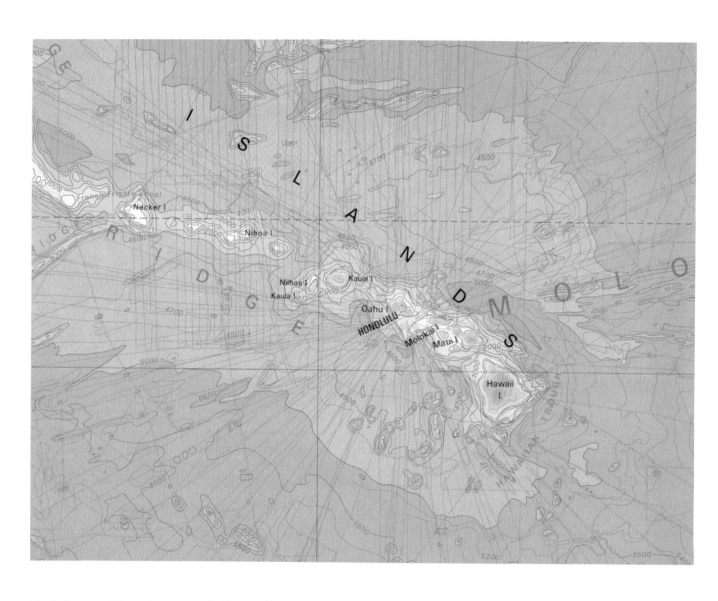

The bathymetry of the sea bottom near the Hawaiian Islands reveals a distinct moat and raised arch, especially near the southeast end of the chain.

However, it was not until 1950 that continuously recorded echograms showed the incredibly obvious results of nature's experiment with loading a thin plate. The sea floor is flexed down into the moat and up into the arch in a simple, smooth curve that resembles a textbook illustration. Curving around the base of the islands in a U tilted to the northwest is a moat about 5500 m deep. Beyond it, 300 km from the center of the line of the islands, is a U-shaped arch. The relief of the arch is about 650 m to

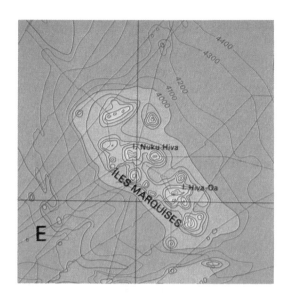

The moat and arch around the Marquesas is completely hidden by an archipelagic apron.

the southeast of the island of Hawaii, and about 1500 m parallel to and east of the Hawaii archipelago. This variation in depth corresponds with theory because the higher arch is due to a linear load, whereas the lower arch is produced by the point load of only the island of Hawaii. The outermost extent of the arch is about 600 km, so the sea floor is demonstrably flexed for such a distance around the Hawaiian Islands.

Presumably, the width of flexing is comparable around other large volcanoes that grow on lithosphere of the same age as that in the Hawaii area. However, there are no other clear demonstrations of the phenomena because they are either obscured by unrelated topography or buried. The effects of topography can be seen off Baja California, where there are many seamounts more than 3000 m high. Even though the area is well surveyed, no broad, low moats or arches can be detected because the relief is dominated by narrow, high abyssal hills.

Abyssal hills obscure the flexing of the sea floor in some regions, but, oddly enough, their absence in other regions is evidence of flexing. All oceanic crust has abyssal hills when it leaves a spreading center, but, around many western Pacific island groups, the sea floor is smoothed by the cover of gigantic archipelagic aprons consisting of erosional and volcanic debris and probably of voluminous abyssal lava flows. They are particularly well developed around the Marquesas and Samoan islands. The load of these great volcanic islands must have flexed the surrounding sea floor, but neither moat nor arch can be seen. However, the whole area between a group of islands and the crest of its surrounding arch would be a gigantic trap for sediment and volcanic rocks. If this trap were filled, the brim should correspond to the crest of the arch. And so it appears to do—the outer edges of the aprons are marked by the emergence of abyssal hills just about 300 km from the centerline of the group of islands. In short, if the moat associated with the Hawaiian Islands were completely filled to the crest of the arch, it would look exactly like the Marquesas archipelagic apron. Instead, the smooth Hawaiian apron has been uniquely warped down into the moat.

BIRTH AND GROWTH OF AN ISLAND

After liquid magma has risen into the base of a volcano, it can stay within the volcano as an intrusion or it can be extruded from the top peak or the flanks. What happens depends in part on the volume of the magma. Small volumes, especially in large volcanoes, tend to cool rapidly and form intrusions. Larger volumes, and those with easy access to the surface, emerge wherever they encounter the least resistance. Consequently, the shape of a volcano is determined by the channeling of magma to the surface. Indeed, D. L. Turcotte has determined that the general shape of submarine volcanoes is a reasonable approximation of a surface that would result from flow through a permeable medium. It is the shape expected if liquid, more or less viscous magma entered the base of a volcano and spread like groundwater to emerge along the path of least resistance. If so, the volcano may grow upward and outward by plastering patches of rock here and there over the whole surface. For some volcanoes, on the other hand, the central conduit that leads to the peak may always be the easiest path to the surface. Thus, their entire volume would be extruded from the summit and none from the flanks.

What ordinarily happens while submarine volcanoes grow upward toward the sea surface is conjectural at present and will remain so until more volcanoes are swath mapped and dredged. The fragmentary evidence now available indicates that both summit and flank eruptions may be important. As Turcotte showed, the general shape of submarine volcanoes supports the importance of flank eruptions. Moreover, dredging and photography on the rift zone on the submarine flank of Hawaii shows that lavas are extruded at all depths on such rifts. The dredged rocks are young and their density increases with depth, so they did not flow down from shallow water. The complex linear shapes of some seamounts show that they grew along such rifts and thus had flank eruptions. Moreover, the flanks of even some basically conical seamounts are sprinkled with a few small, parasitic cones that mark the sites of flank eruptions.

In contrast, some large seamounts are remarkably smooth, simple cones that may have been formed almost entirely by summit eruptions with lava and debris flowing down the flanks. In Hawaii, the great volume of volcanoes emerged very rapidly to form an almost smooth surface with only very minor, secondary relief such as parasitic cones and summit calderas. The small parasitic cones have a chemistry different from the main volcanoes, and they developed only after the main volume extruded. Judging by Hawaii, perhaps the submarine volcanoes that have such a voluminous source that they will grow up to be islands rarely have parasitic cones while they are submarine.

If a volcano has been built from a point at the top of a cone, it is possible to reconstruct its history. Isostatic balance requires that the volcano sinks in some proportion as it grows upward. The balance may be regional or local, but a reasonable and simple approximation is that the compensating root is twice as thick as the mass being compensated. Thus, a volcano with a relief of 3 km must grow upward 9 km because its base subsides 6 km. Granting isostasy, let us consider the growth of a cone-shaped volcano that rises 5 km from a deep sea floor to the sea surface.

The growth of an island in 1,000,000 years.

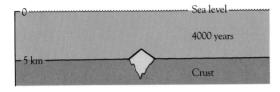

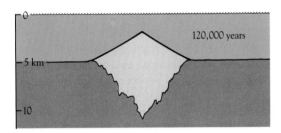

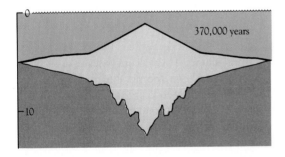

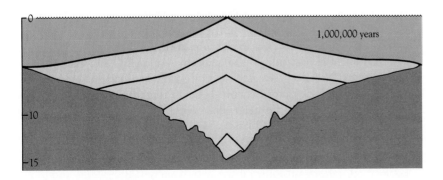

When it is 1 km high, the volcano has only 0.4 percent of its ultimate volume; even when it is 3 km high, the volume is only 12 percent of the future whole. As it grows higher than 3 km, an apron with a slope of only 4° develops around most volcanoes and thus the volume increases even more as upward growth continues. At 4 km height, the volcano has 37 percent of its ultimate volume; and at 4.5 km height, it has 62.8 percent.

On the simple assumption of a constant discharge of magma into the volcano, the volumes can be converted to times. A just-dormant submarine volcano 5 km high has spent 63 percent of its period of activity growing up the last 1000 m and 37 percent of its whole active life growing the last 500 m to the surface. Most of the growth of a Pacific volcano takes less than a million years, before it drifts too far from the source in the mantle. Assuming such a period of activity for a 5 km volcano, it takes 4000 years to become 1 km high, 370,000 years to reach 4 km, and the rest of the time to grow the last 1000 m. It is readily apparent that, if volcanoes that will become high islands are growing randomly in time, at any given time most of them will already have risen up to shallow depths. This is in fact what is observed at present. Occasional eruptions seem to occur at all depths, but the repeated, closely spaced eruptions characteristic of the rapid growth of Pacific volcanoes all take place on relatively shallow peaks.

The abundance of vesicles and thus the density of volcanic rocks vary markedly with depth. Consequently, the density of the rocks in a volcano can be estimated from the assumed mode of growth. For example, if the 5 km volcano grows by plastering on material at all depths, hardly any of the volume is extruded in shallow water and the average density of the rocks is 2.99 gm/cm^3. In contrast, if the volcano grows by summit eruptions, most of the volume is extruded in shallow water. In that case, taking the density-depth relation observed off Hawaii, the average density of the rock in a 5 km volcano should be only 2.61 gm/cm^3. Calculations of den-

Ship Rock, New Mexico, is the remains of an ancient volcano. The volcanic chimney and radial dikes have survived because they are more resistant to erosion than the rest of the volcano.

sity based on measurements of gravity anomalies indicate that some seamounts have a density of only 2.4 gm/cm^3. It has been assumed that the low density of the seamounts is a consequence of large voids between overlapping flows of dense rock, but perhaps vesicles within the rock are also important.

Giant Slumps

The shape of an active submarine volcano ordinarily is a result only of processes of accumulation because, unlike subaerial volcanoes, it is unaffected by erosion. However, above or below sea level, every volcano is affected by gravity, and the consequences sometimes are quite spectacular. As magma enters it, a volcano swells. Radial cracks may develop and sometimes they fill with magma dikes (or the magma pressure forces them open). The solid, crystalline dikes are resistant to erosion, so they, and a volcanic neck, may remain when the bulk of a subaerial volcano has been eroded away. Consequently, the development of radial cracks and dikes is a well-known phenomenon. Less understood is a related phenomenon, the formation of enormous sectoral slumps. The expansion of a volcano apparently can be concentrated in one radial sector, so the cracks bounding the sector become faults with differential movement. Like a giant landslide, the sector slumps downward leaving a gigantic incised valley near the top of a volcano and a bulge below.

The active volcanic island of Goenoeng Api, in Indonesia, displays an excellent example of a sectoral slump. The volcano rises as an isolated and well-formed cone about 3000 m above the floor of the Banda Sea. The perfection of the cone is marred by two very large and one small slump scars on the upper slopes and by fan-shaped slump deposits at corresponding positions at the base of the slope, underwater. Two of the slump scars are subaerial at the top; they have steep, radial side walls and form deep embayments in the small, otherwise circular island.

Similar embayments in other volcanic islands and atolls probably have the same cause, although the evidence is not as obvious. The most spectacular slumps proposed are in the Hawaiian Islands. J. G. Moore has noted that embayments exist in the submarine slope on the north side of Molokai and the east side of the adjacent island of Oahu. Below these embayments, the deep sea floor is quite hilly, unlike the surrounding smooth archipelagic apron. Moore believes that this topography is the result of slumping from the steep upper slopes. If so, the scale is vast. The proposed slump scars are 30 km to 40 km wide and the slump deposits cover an area of more than 10,000 km^2.

Although the enormous submarine slumps proposed by Moore are conjectural, related rift zones on the islands are not. R. S. Fiske and E. D.

A fault scarp on the smooth southern slope of Kilauea.

Jackson modeled gravitational effects by making model "volcanoes" with jello. The jello mounds deformed and rifted on the flanks in characteristic curved lineations. The experimenters then showed that the flanks of the great Hawaiian islands are similarly rifted. Kilauea volcano lies on such a curved rift on the south side of Hawaii. Fault displacement occurs on the rifts just like the movement on a gigantic landslide or slump. Such fault scarps form cliffs that break the smooth southern slopes of Kilauea. However, voluminous lava flows constantly emerge, fill and bury the opening rift and the growing cliffs.

Presumably, adjustment to gravitational effects is almost complete when volcanism ceases. Otherwise most of the flanks of the existing volcanoes would have slumped or slid under the sea.

Volcanism at Sea Level

The results of volcanic eruptions by seamounts just underwater are spectacular, transient, and dangerous and difficult, at best, to observe. Eruptions just at sea level by otherwise submarine volcanoes are similar in most ways but somewhat more visible. Liquid lava at 1000°C encounters cold water, explodes in clouds of steam and ash, and shatters into fragments. Hyaloclastites collect in piles, drifts, and cones and are immediately eroded by waves. Small islands come and go under the influence of volcanic construction, wave erosion, and vertical movement as the volcano swells and deflates. These cataclysms continue, with the construction of a broadening platform, until by chance a lava flow emerges on a transient bank and armors it with basaltic rock. As the next chapter will establish, waves erode basalt volcanoes with surprising speed, so even the first layers of armor are unlikely to endure. However, some volcanoes eventually overpower the peripheral erosion of waves and grow to become islands.

Once a solid, reasonably safe island platform exists, it becomes possible to observe in detail what happens when lava flows enter the ocean. Such observations have been made for decades by members of the Hawaiian Volcano Observatory. The phenomena that occur depend largely upon the type of lava, the rate of discharge, and the intensity of wave action at the shoreline. When blocky aa flows reach the shoreline, they are readily penetrated by seawater, and steam is formed in the hot interiors of the flows. Explosions result, and fragments of lava are hurled in the air in what are called *littoral explosions*. Waves tend to move the small fragments away from where the flow enters the sea; they are deposited in sheltered areas and form black sand beaches. However, the fragments sometimes accumulate in a littoral cone as much as 100 m high on the shore next to the lava flow; occasionally, there are two cones, one on each side of a wide flow. Littoral cones standing above tropical beaches are natural sites for vacationers' hotels on such islands as Maui and Tahiti.

Left: A littoral explosion at the end of a lava tube.
Right: A black sand beach forms from fragments of such explosions.

Clockwise from upper left: A toe from a lava stream enters the sea; cooled by waves, it forms a crust; a crack forms in the crust; through the crack, the toe extends itself farther into the sea.

When fluid pahoehoe flows enter the ocean, the surface of the lava tends to be quenched as glass and shattered into small fragments that accumulate as hyaloclastite deposits. However, lava flows also tend to develop a thin crust, which is a good insulator. The pressure of accumulating lava in that crust produces cracks, lava flows out, and a new crust forms; in this way, a layer of pillow lavas may extend across the shoreline and across the sea floor. On land, pahoehoe lava commonly forms a narrow stream that crusts over to form a lava tube. If the flow of lava is voluminous and rapid at the shoreline, a tube may extend itself underwater like a pipeline. Indeed, littoral lava tubes may form at the shoreline in some circumstances even though there are no feeder tubes, only open streams of lava, on the adjacent shore. An incoming wave may form a crust on a lava stream, during the brief interval between waves the crust thickens, and if everything is just right the next incoming wave does not smash the tube.

Wave erosion is so rapid that most active volcanic islands are ringed by low cliffs. As the latest lava flows pour over the cliffs, a dramatic contest with the waves occurs. Thin lava crusts are shattered and broken up by the waves. Hyaloclastite debris accumulates like a delta, and is deposited in steeply dipping beds, one in front of another, below sea level. On the hyaloclastite delta, subaerial pahoehoe flows accumulate as the shoreline builds outward.

Volumes of hyaloclastites presumably are produced whenever volcanoes are active right at sea level. Thus it appears that an oceanic volcano has four major components in its structure. The base is dominantly pillow lavas, which are increasingly mixed with hyaloclastites near the top of the submarine pile. Above is a layer of littoral hyaloclastites, and on top are subaerial lava flows. The whole is laced with intrusions in the form of dikes, sills, and volcanic plugs. The steeply dipping hyaloclastites deposited on steep volcanic flanks may create a zone of weakness that facilitates the gigantic slumping or gravity-induced faulting observed in the overlying volcanic rock.

Subaerial Volcanism

The volcanoes of the Hawaiian islands are in a dated age sequence, and they are a model for the three stages of volcanism of a midocean volcanic island. Although these stages are known in detail only from studying subaerial volcanoes, there is no reason to believe that they do no apply to at least the larger seamounts that never grew high enough to become islands. The stages are youthful rapid growth, old age, and a brief rebirth; they can happen as easily underwater as above.

The first stage is shield building. Tholeiitic basalt pours out rapidly to build the main volume of the volcano, including the root, submarine base, and island. Typically, subaerial volcanoes resemble shields or inverted soup plates in cross-section. Mauna Loa and Kilauea, in Hawaii, and several of the active volcanoes in the Galapagos are in this stage. The simple, smooth shield is flawed only by steep-sided, circular calderas near or at the summit. This stage commonly lasts only a few hundred thousand years.

In the second stage, old age, the principal flow of magma available from a relatively fixed hot spot is largely cut off. The remainder is a residue of alkalic basalt and more differentiated rocks. Their main effect on the volcano's morphology is to fill the caldera and spread out as a cap over the top of the tholeiitic shield. The lavas of this stage are more viscous than in the first, and they tend to build small cinder cones, which give the smooth shield a warty cover. These are evident at Mauna Kea, Hualalai, and other volcanoes in this stage on the islands of Hawaii, Maui, and Molokai. This stage is normally finished by the time the island is a million years old.

Diamond Head, on Oahu, is a typical small volcano formed midway in the erosion of a great volcanic shield that built an island.

The second stage is followed by deep erosion before the third and last stage, rejuvenation, begins. A possible reason for rejuvenation related to flexural rigidity will be suggested in Chapter 8. The lavas of this stage include alkalic basalts and more differentiated rocks. Their volume is trivial compared with the volcano as a whole, but they form some of the most familiar features of some volcanic islands. The small volcano Diamond Head, rising above the beach at Waikiki, is an example of the dramatic mode, and trivial volume, of volcanism in this stage.

6 *Islands Without Coral*

We have followed the growth of solitary volcanoes from the deep sea floor up into the air to form islands. In the reverse process, they are eroded by stream and wave and eventually are pulled back underwater by thermal subsidence. We may visualize four stages in the life of a volcanic island in water free of coral. A brief period (10^5 years) of submarine growth, a longer (10^6 years) period of both subaerial growth and erosion, then from 10^6 to 10^7 years of subaerial erosion, and from 10^7 to 10^8 years of increasingly deep submergence until the volcano reaches a subduction zone.

During the second stage, volcanism normally overwhelms erosion and is able to build great peaks with smooth slopes like those of Mauna Loa and the Galapagos Islands. Even in this phase, however, erosion may locally be dominant, and the volcano may have to rebuild its slopes repeatedly. Soon after the initial shield-building phase ends, erosion is uncontested, but even then it is much more intense in some places than in others. After prolonged erosion, a minor phase of eruption may occur and it typically fills valleys and buries sea cliffs with thick flows that resist erosion more than the rocks formed earlier. Thus the stage of volcano life with interacting volcanism and erosion is highly complex in detail.

EROSION

The erosion of volcanic islands is influenced by the mineralogy, lithology, and physical properties of their rocks, their size, shape, height, and geographical position, and the intensity and distribution of the winds, clouds, and waves around them. It is only because the islands have so many fea-

Small volcanic cones arise from an islet near icy Heard Island, not far north of Antarctica. The islet has been eroded almost completely down to sea level.

tures in common that the effects of individual variables can be identified. Thus it is that a protective shield of coral is of paramount importance, and, on these grounds, this chapter is separated from the next. With regard to other variables, individual islands are perceptibly more cliffed by waves on the windward side of prevailing seas than on the lee. Likewise, the rate of river erosion is perceptibly faster on the windward side than on the lee and varies perceptibly with elevation—as does rainfall. As to the relation between wave and river erosion, it often depends on the size of an island. Rivers act on the area of an island, which varies as the square of the radius for a circular island. In contrast, waves act only upon the perimeter, which varies only with the radius. Thus, the relative importance of wave erosion decreases with the size of the island—at least while the island is high. If an island persists long enough for rivers to reduce the interior to a low plane, river erosion becomes very slow and wave erosion again becomes dominant.

Weathering

Surface rocks are physically and chemically altered to soil by a complex process called weathering, which varies mainly with mineralogy, temperature, and the balance between rainfall and evaporation. Although all exposed rock surfaces are subject to weathering, the soils in any given place generally are derived from primary weathering somewhere else. This is particularly true of the soils that have nurtured civilizations. These are largely transported by winds and rivers to the plains where they are farmed. Even the soils of islands as remote as Hawaii have a minor content of fine quartz and mica grains blown from the continents. However, most weathering products on islands are indigenous and therefore highly sensitive to local conditions.

In high latitudes and on very high peaks of tropical islands, freezing and thawing may occur in a daily cycle. Ice expands in cracks and is capable of splitting rock into fragments of all sizes. This process has carpeted the crests of Mauna Loa and Mauna Kea, but no lower Hawaiian peaks, with small angular fragments between the higher outcrops from which they are derived. However, mechanical weathering of this sort is minor compared with chemical weathering.

Volcanic islands are composed almost wholly of glass and fine-grained minerals that are more vulnerable to chemical weathering than most continental rocks. The island minerals and their glasses are dominantly olivine, pyroxine, and plagioclase; the more resistant minerals orthoclase and, particularly, quartz are rare or absent. The island rocks react with carbonic and humic acids in rain and groundwater to form both soluble and insoluble decomposition products, but hardly anything remains of the

A dark recent lava flow (about a century old) overlies an older, weathered flow in the Galapagos Islands.

original rock. This is in marked contrast with the weathering of granite, for example, which produces large quantities of sand-sized grains of quartz and feldspar. The quartz grains persist for times on the order of 10^8 or 10^9 years, first as sand and then as sedimentary and metamorphic rocks.

The carbonate compounds resulting from weathering of island basalt are soluble and are carried away by flowing streams or groundwater. Other weathering products include clays, largely kaolinite, silica, and oxides of iron and aluminum. The common red coloring of island soils is the iron oxide hematite. Rarely, the aluminum oxide, bauxite, is concentrated in deposits large enough to be of possible economic interest. Usually the deposits are in areas of very high rainfall, where aluminum oxide is leached from the top layers of the soil. Bauxite is creamy white to pale brown in color; if exposed by erosion, it stands out against the dark red soil. The white color may account for occasional reports of uplifted coral in the high and inaccessible interior mountains of some islands. James Dwight Dana tried to find such a coral reef reported to be in the interior of Tahiti.

Volcanic islands weather into large volumes of soluble carbonates and easily transportable clay but hardly any sand—the very material that makes beaches and river beds. The rivers commonly are paved with rounded cobbles that continue to weather. Black sand is found in small volume on the beaches of active volcanoes, but these sands weather rapidly. The clay and iron and aluminum oxides form soils that are famous for plantations of pineapple and sugar cane. Thus, the weathering products are not all immediately lost to the island. However, island slopes are steep, rainfall high, and erosion rapid, so the soils, from a geological viewpoint, are merely perched briefly on the island before they are carried down to sea level. There they may be deposited as mud in valleys or, as we shall see, continue down to the deep sea.

Glaciers

Glaciation on the 3500-meter peak of Mauna Kea left subtle traces in the form of parallel scratches on rock and also thin moraines of till. Presumably, inactive insular peaks of comparable height have commonly been the site of glaciers even in the tropics. The Balleny Islands, at 66°S latitude in the Pacific, show the effects of high latitude. The islands are less than 1000 m high, but they are almost covered with ice fields and glaciers. Young Island is the largest in area, being about 30 km long and 8 km wide. In a few places, rocky sea cliffs 200 m high are exposed, but the whole interior and much of the coast is covered with glacial ice. Sea cliffs of ice are 100 m to 300 m high.

Bouvet is a small active volcanic island on the crest of the Mid-Atlantic Ridge not far north of Antarctica. Little else is known about it because

Bouvet Island, 54°26′S, 3°24′E.

ice tongues extend to the sea from the interior "plateau," which appears to be a crater filled—temporarily—with ice. Presumably, glaciated active volcanoes are common in very high latitudes.

By far the best known glaciated active volcanic island is Iceland. Although it is only 1000 m to 2000 m high, it is located between 64°N and 66°N latitudes, so it is sprinkled with enormous glaciers. The largest, Vatnajökull, has an area of more than 10^6 km^2. Iceland lies on the intersection of the Mid-Atlantic spreading center and a hot spot, so volcanism is widespread and intense. Occasionally, volcanoes erupt under the glaciers and produce some of the most spectacular natural events known. Entire lakes are melted within glaciers; when they break through one side of the ice, catastrophic floods occur. One in historical times briefly discharged water upon the surrounding countryside at twenty times the rate of the Amazon River.

Glaciers erode rock rapidly, and alpine glaciers produce great relief such as the Matterhorn and Yosemite Valley. Thus it may be assumed that, when and if insular glaciers melt, spectacular relief is exposed to more normal erosion by wave and stream. However, no such conspicuously eroded volcanic islands are known. In part, this reflects the curious fact that few islands of the right height exist in the intermediate latitudes where glaciation might have occurred in Pleistocene time but not now. In part, it may be due to the relative brevity of glacial epochs and the brief existence of volcanic islands.

Even though no examples have been identified, we can speculate on what happens to ancient volcanic islands in high latitudes in glacial periods. Consider Young Island, in the Balleny Islands. At present it is almost entirely covered with glaciers, which protect it from wave erosion but which, presumably, are themselves eroding the interior into spectacular mountains and valleys. The ice has depressed the pre-existing island, which may now be entirely below sea level. What happens if snow continues to accumulate rapidly and thicken the ice, which is only slowly melted by a polar sea? Thermal subsidence of the oceanic lithosphere always tends to pull any island down. Is it possible that when a climatic change or a warmer ocean current eventually melts the ice, the island would be too deep to return to the surface by isostatic rebound? If so, there may be drowned ancient islands in very high latitudes that have never been planed off by waves. Some insular matterhorn may remain to be discovered deep under the polar seas. Certainly, fantastically eroded mountains are encased in coral that protected them as they subsided in warm water. Perhaps ice has played a similar role.

Waves

Waves and swells clearly are very effective in eroding the shores of volcanic islands. Typically, small islands are "iron bound," meaning that they are circled by towering cliffs with great waves breaking at the base. The cliffs must be seen to be appreciated; at Tristan da Cunha, in the South Atlantic, they are 300 m to 1000 m high and nearly vertical for hundreds of meters. For comparison, the Eiffel Tower is about 300 m high. On small and intermediate-sized islands, waves commonly are more effective eroders than streams, so valleys are left hanging and streams enter the sea as waterfalls. On larger islands, streams carve large valleys, and waves are focused on headlands between them. The precipitous narrow ridges between valleys and the great cliffs at the headlands isolate many valleys on exposed coasts of older islands. Such are the circumstances of the Napali coast of Kauai.

Some historical lava flows that reached the shoreline have already been cliffed by waves, so the erosion is demonstrably rapid on a small scale. To measure what can happen on a larger scale and for a longer period, we can turn to Prince Edward Island. It is a small elliptical island, about 5 km by 10 km, that lies at 46°37′S latitude in the southwestern Indian Ocean. Its location in the "roaring forties" exposes it to strong and

The Napali Coast of Kauai.

Reconstruction (color) of the side of Prince Edward Island that was worn away by ceaseless ocean swells. Altitudes are in meters.

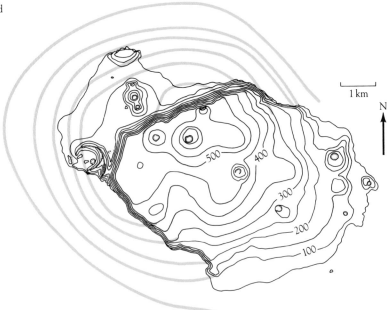

frequent gales, which bring great swells, essentially without interruption, from the west and southwest. About 215,000 years ago, brief volcanism built a broad, gently sloping shield that rose 500 m above sea level. The island is too low and small for much rainfall, so stream erosion has been negligible. However, the great swells truncated the western side of the island to below sea level and carved a cliff 500 m high. About 15,000 years ago, minor volcanism on the insular shelf at the base of the cliffs rebuilt part of the shelf above sea level, but the original shape of the volcanic shield can be reconstructed from a detailed topographic map of the island and surrounding sea floor. The average thickness of rock removed was 172 m, and a shelf 2 km to 3 km wide was cut. It is difficult to make a direct comparison, but the rate of erosion is about the same as in the extreme conditions of the Himalaya Mountains, where, on average, a layer of rock a meter thick is removed every thousand years.

Pitcairn Island, the haven for the mutineers of the *Bounty*, is smaller and older than Prince Edward but not so intensely eroded because it lies in the more equable waters of the subtropical central Pacific. It is 2 km wide by 4 km long, 347 m high, and 450,000 to 930,000 years old. Rainfall averages 200 cm/yr, but stream erosion is negligible because the rocks are permeable. In contrast, the island is ringed by cliffs and there is only one landing—where the mutineers unloaded and burned the *Bounty*. In some places, the cliffs are less than 50 m high, less than a twenty-story building, but 200 m is common. Even so, the cliffs do not compare with those at Prince Edward.

It might be thought that the cliffs of Pitcairn are lower because the island is older than Prince Edward and has sunk along with the sea floor. This idea can be examined by studying the depth of the insular shelf break, which is the outer edge of the gently sloping terrace cut into the original volcano at the same time as the steep cliffs. The terrace is sloping, rather than level, because of Pleistocene and earlier fluctuations in sea level. Numerous samples indicate that most insular shelves are mostly bare rock, and the slope of the volcano flanks below the shelf break has not been oversteepened by sedimentary deposits. Thus it appears that the depth of the shelf break is a reasonable measure of the subsidence of an island. The depth of the shelf break of 43 noncoral islands ranging in age from active volcanoes to 20 Ma shows no systematic trend. Yet in 5, 10, and 20 Ma, young sea floor sinks 700 m to 1500 m; older sea floor sinks more slowly, but even crust 40 Ma old sinks hundreds of meters in 20 Ma. We come to the conclusion that even though the sea floor sinks because of cooling, the islands that are free of coral do not sink. (It will be shown in the next chapter that islands ringed by coral sink rapidly.)

Although there is no correlation between an island's age and the depth of the shelf break, there is an excellent correlation between age and shelf width. This is most readily seen in the Canary Islands, which drift only very slowly because they are near the pole of rotation of the African plate. Thus, the Canary hot spot has been producing large volcanoes for at least 18 Ma, and they overlap to form the existing islands. Tenerife is a triangular island with a large central active volcano, which does not spread to the coast. Most of the coastal rocks are only 0.7 Ma to 1.9 Ma old, but sections of two isolated corners are much older. Wave erosion has been most intense on the northeast corner. There the relation between age and shelf width is consistent, as the table on this page shows. The shelf also changes abruptly in width at the boundary between rocks of different age on the southwestern, leeward side.

If the ages and shelf widths of eight isolated Atlantic islands and eleven major sectors of the Canary Islands are compared, the simplest interpretation of the observations is that waves have widened the shelves by 0.6 km/Ma to 0.7 km/Ma for at least 16 Ma. A similar analysis for the relatively reefless Hawaiian and Marquesas islands indicates widening by 1.1 km/Ma to 1.7 km/Ma for 5 Ma to 7 Ma. If all dated volcanic islands, including active volcanoes, are considered together, the relation between shelf width and age does not appear linear. Many islands, such as Prince

Age of rocks and width of the insular shelf on the northeastern coast of Tenerife

Age (Ma)	Width (km)
0.7–1.4	1–3
5–7	3–3.5
15.7	4.5–5.5

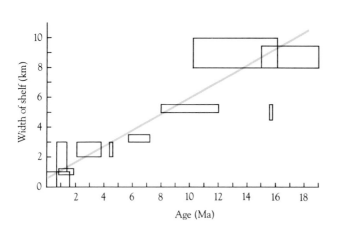

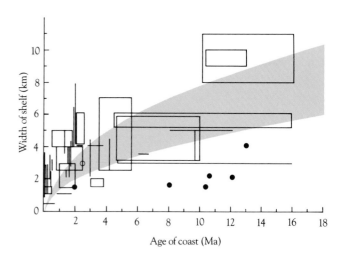

Left: The regular relation of age to the width of the shelf break for eleven major coasts in the Canary Islands.
Right: Extended to many coasts in the Atlantic, Pacific, and Indian oceans, the relation is less regular.

Edward, that are less than a million years old have shelves 2 km to 3 km wide so initial erosion appears to be particularly rapid. This seems reasonable in that waves lose their power to erode when they lose energy by friction with a widening shelf. However, the age-width data are relatively scattered for older islands, which suggests that various other factors become increasingly important.

One such factor is certainly the intensity of swell, as can be seen by comparing different sides of the same island. For example, east of Australia is Lord Howe Island, a later haven of the refugee descendants of the *Bounty* mutineers. It is a small island rising from a broad shelf but not from the middle of it. The island is 14 km from the shelf edge on the weather side and only 5 km from the edge on the lee side. Youthful Ascension Island, in the Atlantic, shows a similar relation, with a weather shelf 3.5 km wide and a lee shelf only 1 km wide. Much older (19 Ma) Gomera Island in the Canaries shows a greater gap: weather side 9 km wide, lee side 3 km wide. It is interesting to consider the wave-cut platform that will remain when one of these islands is at last eroded away and sinks to become a guyot. The truncated top of the guyot will not be flat, nor will it be a right cone sloping uniformly from the perimeter to a point center. Instead, it will be a cone with the apex off center and closer to the former lee than the former weather side of the island. Thus, in favorable circumstances, it may be possible to determine the direction of the dominant swell that truncated a guyot that has drifted into an entirely different latitude since it was an island.

Lord Howe Island, east of Australia, sits well off center
on its shelf. The isolated pinnacle in the distance is Ball's
Pyramid (see page 125).

Rivers

The principal factors that influence the erosion of a volcanic island by
rivers are the permeability of rock, the shape and slope of the island, and
the quantity and distribution of rainfall. Permeability is important because
fresh lava flows are highly permeable, and rainfall may simply penetrate to
become groundwater instead of flowing off as surface streams. Most permeable are flow surfaces, where gas bubbles accumulate, so the thin flows of
the shield-building stage are particularly permeable. As time passes,
weathering produces relatively impermeable clay soils at the surface, and
deposition of minerals from groundwater tends to fill openings and reduce
porosity and permeability. Thus it is that the ratio of surface runoff to
groundwater flow gradually increases in accord with the age sequence in
the Hawaiian Islands, from 0.8 in the young island of Hawaii to 3.8 in the
old island of Kauai. The only exception is Oahu, where groundwater flow
exceeds runoff.

The residence time of Oahu groundwater is decades to centuries, so
the volume of groundwater is enormously greater than the amount of
surface water. In fact, almost any oceanic island, high volcano or atoll,
contains a vast reservoir of fresh water. Islands are so permeable that

The Ghyben-Hertzberg lens of freshwater (light blue) under an oceanic island rests on denser saltwater (dark blue) that permeates the base of the island.

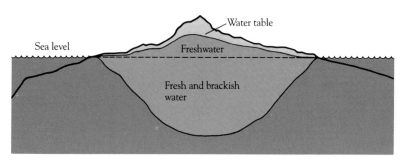

seawater saturates them and moves with the tides, although with some time lag. In the absence of rain, the seawater within an island is at sea level. However, rainwater percolates through the island and floats on the seawater in what is called a *Ghyben-Herzberg lens*. The lens of fresh groundwater forms because flowing groundwater has a slope, from the interior to the coast, that is inversely related to permeability. Because of hydrostatic balance, the lens is biconvex. Like the island itself, the groundwater has a root; for normal densities of fresh water (1.0) and sea water (1.025), the root is about 40 times as thick below sea level as the water it supports is above sea level.

Volcanic islands generally are circular or elliptical cones or domes, and it is easy to visualize the influence of their shape upon erosion by imagining simple circular cones that lie in seas without waves and on which rain falls uniformly. The consequent rivers that develop on a cone are radial because the slopes of the cone are radial. The side slopes of the river valleys tend to be relatively constant but the longitudinal slopes are steeper in the headwaters than at the shoreline. Thus the valleys of the radial streams are funnel shaped; they are narrow and shallow at the shoreline and spread into great, deep amphitheaters in the interior. At the shoreline, the rivers are more widely spaced than in the interior, and between them the conical volcanic slopes appear relatively unmodified; these are such distinctive features as to have a special name—*planeze*.

The planezes have a triangular plan because, as time passes, the widening valleys converge in the interior and are separated only by knife-edge, steep-walled ridges. These ridges become so narrow that, in such places as Moorea and the Marquesas Islands, there are great natural windows that cut through them. Farther in the interior, the amphitheaters overlap, and the whole interior of an island may be gutted while the peripheral planezes are relatively intact. The formation of an erosional caldera by this process has been so rapid that it has removed the central peak of Tahiti Iti, the southern peninsula of Tahiti, which is only 0.4 Ma old. The center of the older main cone of Tahiti (0.5 Ma to 0.9 Ma old) is

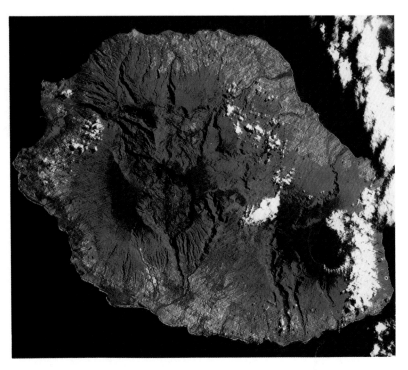

Left: The relatively uneroded planezes between deep gullies are preferred locations for human settlement on Reunion. *Right:* A satellite view of Reunion shows the radial pattern of erosion and the gutted interior of the island.

also an erosional caldera. If erosion by waves is added to this picture, the planeze headlands are cliffed between the rivers. Consequently, the coastal apices of the triangular valleys are also removed, and the coastal valleys have more parallel sides.

All the simple phenomena described above have been observed, but the fluvial erosion of some volcanoes is more complex because rainfall is not uniformly distributed. Several effects are important in the distribution of rainfall. One is orographic. A moving airmass encounters a mountain, is forced upward and cooled. The dew point drops, and rain or snow falls on the mountain crest. In the Canary Islands, winds come from many directions and 120 cm of rain a year falls on the 2000-meter central peak of Gran Canaria Island, but only 20 cm falls along the shore. However, the pattern can be different if almost all of the rain comes from one direction, as it does in the Hawaiian Islands. There the northeast trade

A knife-edged ridge on Moorea.

winds are laden with water up to about 2000 m and they flow in complex patterns around and over the high islands. Circular, relatively low-lying Kauai is entirely within the humid air mass and the rainfall is closely related to elevation, ranging up to 1100 cm/yr at the peak, in a pattern comparable to that of Gran Canaria. Hawaii is much higher than Kauai, and the rain-filled clouds pass around the peaks, so most of the shoreline is in a soggy belt with 250 cm/yr to 750 cm/yr of rain, while the peaks are deserts with only 40 cm/yr. The morphological effects are not very great on the slopes of the active or very young volcanoes, because the rain penetrates to become groundwater and new lava flows fill incipient valleys. However, the northern tip of the island is Kohala Mountain, which has been extinct for 400,000 years. It is about 2700 m high and elongate northwest-southeast and therefore perpendicular to the trade winds. Rainfall somewhat east of the peak has a high of 320 cm/yr and it is about 220 cm/yr on the windward coast. On the lee coast, it is only about 25 cm/yr. The dissection of the once smooth volcano has an almost uncanny correlation with the rainfall. Directly under the most intense rainfall are the headwaters of three great valleys. To the northwest on the weather coast, where rainfall is much less intense, are many small gullies. In contrast, the leeward desert is virtually undissected.

It takes little imagination, now that the potassium-argon ages of the islands are known, to see a close correlation between age and the degree of

The water-laden northeast trade winds give rise to rain clouds mainly near the tops of mountain ridges on Oahu.

dissection, after allowances for patterns of rainfall. However, Chester K. Wentworth in 1927 performed the much more difficult feat of calculating the age of the islands from the degree of dissection alone. Wentworth reconstructed the original shapes and volumes of the Hawaiian volcanoes and compared them with the present volumes, as determined with a planimeter and topographic maps. Dividing the eroded volume by area, he obtained an average depth of erosion that may be compared with the ages as now known: Hawaii 8.7 m (active), Kohala 13.6 m (0.4 Ma), Maui 40 m (1.0 Ma), Molokai 70 m (1.6 Ma), Oahu 100 m (2.7 Ma), and Kauai 127 m (4.4 Ma). A smooth concave-upward curve correlates age and depth of erosion to 1.6 Ma. I have recalculated the average rate of erosion for Kauai using much more detailed maps than those available to Wentworth. The depth of erosion on Kauai appears to average 271 m, which gives a smooth, concave-upward curve for at least 4.4 Ma. If, instead, the rate of erosion is calculated it appears to decline in accord with standard observations in continental geomorphology. Comparing windward sides of high islands, the average rate has declined from 280 m/Ma

The relation of age (left) and erosional depth (right) to distance from Kilauea.

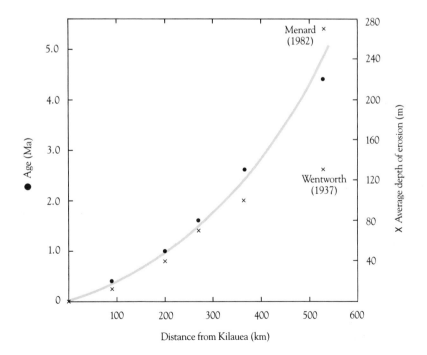

for young Kohala to 40 m/Ma in older Oahu. In contrast, the rate on the leeward side of high islands and all sides of low islands is only 2 m/Ma to 6 m/Ma and hardly varies at all with age. The average rates of fluviatile erosion of these islands can be compared directly with that of the Himalayas, which is about 1000 m/Ma. It is evident that the spectacular relief of volcanic islands is not a consequence of rapid erosion but of the peculiar conical shape and structure of the volcanoes.

In order for Wentworth to convert his erosion data to rates and thereby determine the age of the islands, he needed a historical determination of an erosion rate. This he found on the island of Lanai. The kiawe tree had been introduced to Hawaii in 1837, and the trunks of older trees were buried to a depth of as much as a meter on the narrow coastal plain. Knowing the volume and age of sediment and the area of the source region, Wentworth determined an erosion rate. From it, he obtained an age of Kohala of 225,000 years and of Kauai of 2,090,000 years. These are in just the right proportion but are half the ages measured by modern isotopic methods forty years later. Wentworth's measurements were soon criticized and then largely ignored for decades, despite the fact that they could have been very useful if accepted.

Marine Dispersal of Sediment

The solid rock of volcanoes is converted by weathering to soluble carbonates and silica that disappear into the ocean. Much of the clay likewise is dispersed, but in suspension instead of solution. However, a minor residue of mud, sand, and gravel passes on to the insular shelves. Where is it finally deposited? Seismic profiles of the bottom beneath the sediment indicate that sediment does not form thick deposits on the shelf or the deeper slopes. This confirms the thousands of notations on older nautical charts that record rocky bottom around islands. A detailed survey around Oahu in 1974, for example, showed that sediment on the shelf was mostly less than 10 m thick. The only exceptions were small accumulations along the bases of drowned sea cliffs and river valleys. It appears from the distribution that the sediment is merely being transported along the bottom toward the deep sea rather than accumulating in place on the shelf. However, the picture, as always, is clouded by Pleistocene changes in sea level. Perhaps it is more significant that sediment does not accumulate on the deep flanks of islands. The fraction that is not dissolved or carried away in suspension apparently is mainly transported to the deep sea bottom beyond the submarine slopes of the islands. There, in fact, it is readily detected because it forms an apron and buries the surrounding abyssal hills. Deep-sea cores show graded bedding and other evidences that the sediment flows rapidly from shallow water to deep in the form of turbidity currents. In sum, hardly any of the products of weathering and fluviatile and wave erosion remain on the volcano from which they are derived.

Isostatic Uplift in Response to Erosion

Two lines of evidence indicate that islands in cool waters do not subside. One of these is that the depth of the shelf break does not increase with age but the width of insular shelves does. The only explanation seems to be that the shelves, and thus the islands, remain at about the same level and are progressively eroded away by waves.

The second line of evidence is that, in the interiors of some islands, rock formations are exposed that appear to have been deposited under water. The evidence at La Palma in the Canary Islands is particularly convincing: the formation includes pillow lavas, breccias containing fragments of pillows, and hyaloclastites. A slaggy lava formation on St. Helena also appears to be submarine in origin. The exposure of these once submarine rocks has involved local tectonic elevation and tilting. However, the point of interest here is that submarine rocks are not necessarily buried deep under the subaerial lava flows of an island that has subsided because of loading and also cooling of the deep sea floor. On the contrary, the submarine pedestal of the center of some islands has remained near sea level, and even been uplifted and exposed by erosion.

If these interpretations are correct, an anomaly appears. The deep sea floor sinks, barrier reefs and atolls sink, guyots sink, but volcanic islands that are free of coral do not sink. The only distinctions of such islands is that they alone are exposed to wave erosion and lose their weathering and erosional products to the ocean and the deep sea floor. Thus, it appears that these islands stay at sea level because of isostatic rebound as they are eroded away. Likewise, it appears that the persistence of islands depends on their size. They endure until they are eroded away. Then they sink at the same rate as the cooling lithosphere on which they rest.

SUBMERGENCE

In general, thermal subsidence is balanced by erosion-isostasy, but in some circumstances it may only be slowed. For example, the thermal subsidence of very young lithosphere may be too fast to be balanced by the uplift of a small low island that is not eroded by streams. A combination of wave erosion and subsidence may produce either a series of narrow terraces separated by cliffs, or, less likely, a smooth but relatively steeply sloping shelf. There are few places where the existence of such topography can be confirmed. The only small, isolated volcano known to have grown upward fast enough to be an island next to a midocean ridge crest is now the submerged Cobb Bank off Washington. It is 1.5 Ma old and lies on crust 3 Ma old, so it was active on crust less than 1.5 Ma old. Cobb has sub-merged terraces at two general levels, 823 m to 1189 m and 183 m to 200 m, as well as a broad summit platform at 82 m and a single central pinnacle rising to 34 m. It appears that a simple, small conical island was notched by waves and almost truncated as it rapidly subsided. Rounded cobbles on the terraces support the interpretation that they were once beaches. Thus Cobb Bank has the morphology predicted for a small island on very young crust.

Pinnacles, Rocks, and Banks

In the slowly drifting parts of the Atlantic, high volcanic islands persist for long periods because of renewed volcanism. Either sources of magma drift with the islands, or, more probably, relatively fixed hot spots can be tapped from some distance. In contrast, islands on the rapidly drifting Pacific plate ordinarily have only one major shield-building phase and one minor phase a few million years later. Hardly any volcanic islands in the Pacific are more than 10 Ma old and most are less than 5 Ma. Thus, it appears that, without repeated volcanism, even large islands are eroded to near sea level in 5 Ma to 10 Ma. What then remain are pinnacles, rocks, and banks with elevations of about +300 m to −200 m. The pinnacles and rocks are fearsome spires rising almost vertically 50 m to 300 m above the sea. Most are isolated, although some are flanked by a separate smaller rock or two. They rise from extensive and relatively flat rocky platforms and clearly are the last erosional remnants of once extensive islands. Banks, in nautical terms, are defined as features that do not break the surface and are less than 200 m deep. Some of them, such as Cobb Bank, preserve a small central spire that somehow escaped final truncation by waves. Most, however, are relatively smooth, showing that truncation was complete.

Banks are in the euphotic zone, and biological productivity tends to be high, but fine-grained organic remains generally are stirred by waves

Ball's Pyramid, not far from Lord Howe's Island, is the last subaerial remnant of an eroded oceanic island.

and swept away by currents. Thus, most of the surface of a bank is bare rock. Once below the depth of vigorous erosion, the mass of a bank remains constant and there is no isostatic rebound to counter thermal subsidence. If sea level were constant, all banks thus would have a simple subsidence history. However, sea level fluctuates; with every large eustatic lowering, banks are reexposed and erosion-isostasy acts again. Consequently, the actual history of banks is longer and more complex. Still, thermal subsidence is persistent, and banks that do not drift into coral seas sink eventually beyond 200 m depth and the range of eustatic fluctuations. The average time that a bank exists depends on the age or thermally rejuvenated age of the underlying lithosphere. If an island existed for 10 Ma before it was truncated to a bank, the lithosphere is at least 10 Ma old. Lithosphere of that age subsides through the depth range of banks in only 3 Ma to 4 Ma. However, a bank on lithosphere 100 Ma old may persist for 8 Ma to 12 Ma before it sinks below 200 m and, by definition, becomes a guyot. In that case, the bank is in the depth range for eustatic change in sea level for a longer period. Thus it is hardly surprising that fossils derived from shallow banks are common in deep-sea drilling cores in the Pacific, nor that they were deposited at times of lowered sea levels.

Guyots

The banks that at last escape below 200 m become guyots by definition, but except in depth and name they are unchanged at first. Thereafter, in the average lifespan of 10^8 years, little can happen to a guyot to change its appearance. Gravel has been dropped on summit platforms by Pleistocene icebergs in the Gulf of Alaska and on the Emperor Guyots, but the deposit appears to be thin and uniform. In exactly the right circumstances, however, the platform shape is modified. In the South Pacific, many guyots sank to depths of hundreds of meters before they drifted north into the warm seas that would permit coral to convert them to atolls. Some of these guyots have drifted farther north under the zone of high productivity in equatorial surface waters. Consequently, their summit platforms are covered with hundreds of meters of layered pelagic sediment. This smooth pelagic cap is thickest in the middle of the platform and slopes toward the edge. The mere existence of a thick cap indicates that a guyot in the Northern Hemisphere was created in the Southern. By sampling the cap, the time when the guyot crossed the equator can be determined as well.

The only other thing that happens to guyots is that some bob up and down in addition to their normal thermal subsidence. This bobbing amounts to hundreds of meters in tens of millions of years. Some upward and downward motion may be due to distortions of the lithosphere by convection currents in the mantle; if so, they have not been detected. In contrast, it appears probable that guyots are elevated as they override the Tahitian midplate swell and presumably others. Certainly, volcanic islands and guyots develop on or near the crest of the East Pacific Rise. The guyots on the rise flanks east of the Tahitian swell should have sunk with the lithosphere and thus be at depths of 1500 m to 2000 m by the time they reach the swell. However, the guyots on the swell are all at depths of less than 1000 m.

Guyots must also be uplifted near subduction zones as the lithosphere curves upward, because of its rigidity, before it plunges into the mantle. This can be shown to occur for atolls, as will be discussed in Chapter 9, but it is difficult to document for guyots. However, the age sequence of the Pratt-Welker guyots in the Gulf of Alaska is quite well known, and the guyot on the outer swell of the Aleutian Trench is shallower than it should be for its age. Presumably, it has been uplifted. There are also guyots in the trenches of the western Pacific, and some of them have been pulled down thousands of meters along with the sea floor. What then happens in subduction zones will be left hanging (until Chapter 9) while we return to the open ocean.

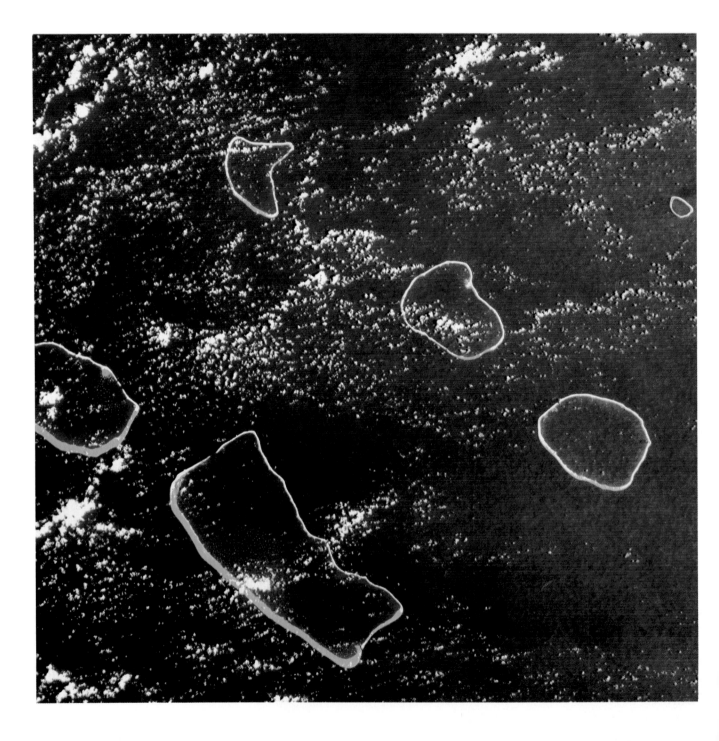

7

Islands with Coral

Charles Darwin had never seen an atoll when he conceived of a hypothesis for their origin. However, he was on a surveying ship, H.M.S. *Beagle*, surrounded by charts and descriptions of different types of islands, and that was enough. There were already explanations for atolls, because explorer's descriptions of great rings of wave-girt coral around calm lagoons in the deep Pacific had captured the interest of stay-at-home scientists. Darwin's mentor, Charles Lyell, had proposed that what he called "lagoon islands" were "nothing more than the crests of submarine volcanoes having the rims and bottoms of their craters overgrown by coral." Other hypotheses, then and after, were equally ad hoc. Darwin's hypothesis, in contrast, was a powerful synthesis that explained many phenomena previously thought to be unrelated. Darwin began by noting that there were only three types of oceanic islands, namely, volcanic, coral, and combinations of the two. He simply proposed that there was only one sequence of development of all oceanic islands. Volcanoes grow up from the sea floor to form high islands and then die. Coral grows in the shallow water fringing the shoreline. The volcano subsides and the coral grows upward, leaving a gap between a barrier reef and a central volcanic island. The subsidence continues until the volcanic island disappears and nothing remains but an atoll.

The hypothesis in itself was attractive for its simplicity and its synthesizing power, but Darwin did not leave it at that. He set out to confront all the critical points in the hypothesis with data already available. It appears that, by the time he had developed the idea, the formation of an empty lagoon encircled by coral seemed rather obvious—if volcanoes subside. He

Atolls in the Tuamotu archipelago. Only coral is exposed, but it rests on volcanoes that were active along narrow ridges roughly forty million years ago.

therefore concentrated on the proof of subsidence. This he derived from proving the following:

1   Corals do not live below a certain depth (d).
2   Coral reefs may be thicker than d.
3   Coral reefs can grow upward fast enough to remain in shallow water as the underlying sea floor subsides.

As to the depth at which corals grow, he had made observations using a sounding line with a bell-shaped weight full of wax to which loose bottom samples adhered or which showed the shape of hard bottom. Above 36 m, he found coral; below, sand. He visualized the boundary as one determined by a variable environment and cited observations, by others, of deeper coral. However, nothing supported the existence of living coral below 56 m.

Concerning the thickness of reefs, the island of Mangaia, east of (and uplifted by the load of) Tahiti, is an uplifted atoll about 110 m high. Other atolls were uplifted enough to expose coral more than 56 m thick. Clearly they had subsided before they were elevated.

As to the rate of growth of coral, the general belief in Darwin's time was that corals grow extremely slowly, because some in the Red Sea were believed to be unchanged since the time of the pharaohs. Moreover, the depth of the living coral near the entry to Tahiti harbor had not changed from 4.5 m in 67 years of sounding by British ships. Darwin reasoned that the subject had not "hitherto been considered under a right point of view." A coral reef is alive, and like all living things its growth is controlled in part by the environment. The proper question is not how rapidly a reef grows in unfavorable conditions, but how rapidly it can grow in favorable ones, as when an island subsides. He was able to cite experiments and accurately timed observations of fouling of ship bottoms. They showed that coral can grow 360 mm/yr and possibly 1560 mm/yr, although Darwin himself doubted the latter number. In any event, he believed that coral would have no difficulty in keeping up with subsidence due to any geological cause known to him. However, he did not know about the speed of Pleistocene changes in sea level.

Darwin's hypothesis was published in 1842 and was generally acclaimed for decades as a triumph of scientific logic. The great Lyell discarded his own hypothesis and supported that of his protege. Darwin had become famous as a consequence of his correspondence and his readable books about his experiences on the *Beagle*. His hypothesis not only had

Top to bottom: Moorea, Bora Bora (both in the Society Islands), and Aratika (in the Tuamotus) represent successive stages in the transformation of a subsiding volcanic island into a coral atoll.

great appeal in itself, but was based on the personal observation of the most qualified scientist in the world. If possible, acceptance of the hypothesis was even stronger and more widespread after 1872, when Dana published his book *Corals and Coral Islands*. After reading the newspaper account of Darwin's ideas in Sydney in 1839, he had sailed to the Fiji Islands and there seen "similar facts on a still grander scale and of more diversified character." Dana remarked that he was more positive than Darwin about the subsidence hypothesis. Moreover, Dana showed that features of islands that he had observed, but that were unknown to Darwin, could have been predicted from the basic hypothesis. Thus, the two leading observers of islands on early nineteenth-century scientific expeditions were in complete agreement.

Critics of Darwin's Hypothesis

However, new types of observations are the commonplace of science, and, in the second half of the century, the *Challenger* Expedition set out to explore not the islands but the sea between them. John Murray, the only geologist, found that the skeletons of the calcareous microorganisms that lived in surface waters accumulated on the deep sea floor at depths from about 2500 m to 4500 m but were absent at greater depths. He and his colleagues concluded that the skeletons dissolved in the cold deep water. Drawing on his expertise in deep-sea sedimentation, Murray rejected the subsidence hypothesis in April 1880. Atolls were formed because sediment accumulated on undersea mountains and eventually produced shallow banks. When a bank was shallow enough, corals attached themselves and grew up to sea level. Solution of the coral skeletons in very shallow water produced a central lagoon, although the very favorable environment at the edge of the bank preserved a ring of living coral.

Darwin himself had considered that atolls might develop on great banks of sediment, but he rejected the possibility of such banks in the pure blue waters of the central Pacific. He does not seem to have considered the possibility of sediment accumulating on the peak of a high submarine volcano, as it appears that Murray did. Darwin knew from soundings by *Beagle* around islands that shallow submarine banks usually are swept free of pelagic sediment by waves and currents. In any event, Murray seemed to have a point. Why should all oceanic volcanoes grow high enough to be islands? Indeed, even the widely spaced soundings of *Challenger* showed that the sea floor had submarine mountains of some sort.

Regrettably, Murray's logic failed, and he believed that if atolls could form as he proposed, then Darwin's subsidence hypothesis was wrong.

A coral drill core (actual size).

Many weaknesses in Murray's hypothesis were soon identified and, in any event, it did not eliminate Darwin's ideas or give an alternative explanation for Darwin's and Dana's evidence of subsidence. At best, it appeared to identify a minor variation on Darwin's theme. Nonetheless, Alexander Agassiz, another oceanographer, almost immediately supported it and rejected subsidence. So did Edward Forbes, in his book about his work at Keeling Atoll, published in 1885. H. B. Guppy studied that same atoll and agreed with Forbes in 1888; he also mapped the reefs of the Solomon Islands and concluded in 1884 that barrier reefs could build up from gently sloping shelves without subsidence.

Somehow, perhaps because of the great fanfare regarding the *Challenger* Expedition, Murray's idea caught on from the beginning, just as Darwin's had when *Beagle* returned. Darwin wrote to Alexander Agassiz in May 1881 to ask for his opinion on "Mr. Murray's views." He himself was in no way committed to anything except whatever hypothesis best explained the scientific facts. "If I am wrong, the sooner I am knocked on the head and annihilated so much the better." Darwin wished that some "doubly rich millionaire" would pay for borings of Pacific and Indian atolls to obtain cores down to "500 or 600 feet."

After Murray's ideas gained widespread support, a test by drilling seemed ever more desireable. In 1890, Professor W. J. Sollas in England began to correspond with Professor Stuart Anderson at the University of Sydney. Anderson was to try to borrow a drilling rig from the Australian Government. Sollas whipped up support for the project and proposed it formally at a meeting of the British Association for the Advancement of Science in 1893. A committee of distinguished scientists was immediately appointed to proceed with the project. Eventually it would include a number of British geologists as well as John Murray and Francis Darwin, the son of Charles, apparently to balance matters. In 1894, £10 was allotted by the Royal Society for the annual expenses of the committee. In 1895, the Pacific atoll of Funafuti was selected for drilling. It was one of the atolls closest to the drilling rig, which was already available in Sydney.

In 1896, the Royal Society obtained a grant of £800 from the Treasury and the use of H.M.S. *Penguin* from the Admiralty. Sollas was appointed leader of the small scientific party. The formal objective was to investigate the depth and structure of a coral reef. If the drilling encountered a rock other than coral limestone, Sollas could "use his discretion." Darwin's supporters wanted the possibility of reaching a volcanic platform to be acknowledged. Off Sollas went, only to learn the great difficulties of drilling in reef limestone full of frangible and hard limestone with occasional caverns and cavities. Sollas tried drilling at the lagoon margin but could not spud into the loose sand. After Herculean efforts, he managed to penetrate 105 feet at a site on the seaward edge of the island. The section was all limestone and dolomite and included both coral fragments and the foraminiferal tests that make up the sand on lagoon beaches.

Francis Darwin was a physicist and astronomer, and perhaps it was at his suggestion that a magnetic survey was made of the atoll. Inclination and declination were measured along the ring of islands circling the lagoon. The bathymetry of the lagoon was being surveyed by boat, and a raft was constructed to provide a relatively stable, and iron-free, platform for magnetic observations. The point of the survey was that coral limestone and the platform of pelagic mud proposed by Murray would contain no magnetic minerals and thus mapping would merely show the uniform magnetic field of the Earth at that spot. On the other hand, the basaltic rock of volcanoes does contain magnetic minerals. If Charles Darwin was right, the platform under the coral had the fantastically eroded shape of a modern volcanic island surrounded by a barrier reef. Thus, a magnetic map at the surface of the atoll would be influenced by the variable distance to the basalt not too far below. In fact, the magnetic survey showed that the field was not smooth; instead, there were large "disturbances" localized on the east and west sides of the atoll. It was proposed that any subsequent drilling might be directed toward the magnetic highs with the hope of reaching the material causing the disturbances.

The critical depth of 500 feet hoped for by Darwin had not been reached, so the test was deemed unsatisfactory. In 1897, a second expedition drilled to 698 feet, but the continuing bathymetric surveys of the atoll were interpreted as indicating that the basement platform was only a little deeper. The Royal Society launched a third and last expedition in 1898. It was led by Professor T. W. David; by that time the whole endeavor must have seemed routine and even a bit of a lark, because Mrs. David came along. Lark or not, the atoll was successfully drilled and partly cored to 1114 feet; it was composed of reef and lagoon limestone from top to bottom. Sixty years later, British geophysicists would show that the volcanic basement was below 3000 feet.

After all that effort, the nicely balanced, scientifically neutral committee issued a monograph in 1904 summarizing the field observations. As to their meaning, "Into the controversies about the development of coral reefs, those who have been concerned in the preparation of this volume have not attempted to enter." One of the committee members, Archibald Geikie, had already written in 1903 that Darwin's "simple and luminous" hypothesis was generally accepted by geologists, but in fact it was not. The views of another committee member, Murray, were unswayed by the drilling. Many others argued that the holes were drilled in reef talus that had buried the side of the atoll. Consequently, a basement platform under the lagoon might be quite shallow and composed of any material. The magnetic survey was ignored.

The principal competitors to Darwin's ideas after the Funafuti drilling can all be grouped under what came to be called the "antecedent platform theory." The basic problem of the idea was to account for the existence of hundreds of shallow banks at just the right depth for coral in the tropics. Darwin would have required an equal number of equally shallow banks in temperate zones before being satisfied, but only one of the advocates of this so-called theory—Reginald Daly—was a great generalizer.

Alexander Agassiz, the indefatigable observer of coral reefs and great friend of Murray, offered one version of the theory in 1899. He observed that the fringing-barrier reef on the northeast coast of Tahiti had basalt pinnacles projecting through it. Clearly, he thought, it rested on a largely truncated insular shelf. In this he was correct—the headland cliffs that remain next to the shelf are still visible. He observed that storm waves cross incomplete reefs in some places and attack the rocky shore. He saw no reason why this erosion should not continue until the whole island was truncated. The powerful scouring action of great waves would carry the erosional debris out through the narrow reef passes or over the reef itself. Moreover, the scouring would not stop at sea level but would erode solid basalt down to lagoon depths, despite the encircling reef. All these extraordinary phenomena were proposed to escape any taint of subsidence in the disappearance of volcanic islands.

The idea of wave erosion within a lagoon did not find widespread acceptance. However, the erosive powers of waves on coasts not protected by reefs was beyond question. Thus, several scientists proposed that reef coral did not colonize the volcanic islands that formed atoll platforms until after they were truncated to shallow banks. Just why corals now fringe and encircle volcanic islands but did not when the atoll platforms were truncated posed some problems. A brilliant, although incorrect solution was advanced in 1910 by Reginald Daly. He had been thinking about the marine effects of Pleistocene glaciation and realized that the surface waters would be chilled as well as lowered by a hundred meters or so. There was no way to measure the lowering accurately at that time, but the approximate volume of continental glaciers could be estimated. He began his analysis of the origin of atolls with the observation that the lagoons of the larger atolls have a uniform depth of 70 m to 90 m. How he could have been led to such an inaccurate observation is inexplicable—the depths, ignoring pinnacles, of large atolls actually range from 20 to 90 meters. In any event, he concluded that such uniformity could not be a consequence of sedimentation and, by elimination, must reflect erosion. The cold ice-age waters must have killed or weakened the tropical reef builders and thereby exposed all coasts to erosion. The future atolls were all planed off by waves during periods of lowered sea level. Then the earth warmed, the corals thrived, and all existing atolls grew upward with sea level as the glaciers melted. Barrier reefs grew up from the outer edges of insular shelves of islands that were only partially truncated.

Daly's glacial-control hypothesis was questioned by his colleague at Harvard, William Morris Davis, who supported Darwin's views. It was universally recognized that Daly had identified an important geological factor that had previously been neglected, but those who believed in antecedent platforms continued to seek causes other than glacial control. One group of eminent observers of coral reefs noted that many reefs are elevated. They proposed that antecedent platforms of differing initial depths could all approach sea level from below. Inactive volcanic seamounts, for example, might be elevated to shallow depths where reef corals would colonize them and grow rapidly to sea level. Meanwhile, elevation would continue without interruption and the reefs would die and appear as they do today. Moreover, the seamount might have a coating of limestone from deep-sea sediments as proposed by Murray. If so, the thin reef coral would be lodged on older and possibly thicker limestone that was not of coral origin.

In 1887, Guppy came to the conclusion that reefs, including atolls, were characteristic of regions of elevation instead of subsidence. The atolls simply had not time to rise above sea level. Alexander Agassiz made the same proposal early in this century. In the 1930s, the leading American geological experts on living and recent reefs were J. Edward Hoffmeister and Harry S. Ladd. They had spent three field seasons on the islands in the South Pacific. They found little to support either subsidence or glacial control of reefs. They observed that some uplifted coral reefs rested on foundations of non-coralliferous limestone. They also noted that some so-called elevated atolls had actually been flat banks of limestone before uplift and that the lagoon-like central basin was caused by subaerial solution. They concluded that atolls developed on antecedent platforms having various origins, including uplift—but not subsidence. Guppy worked in the Solomon Islands, Agassiz around Fiji, and Hoffmeister and Ladd from Fiji to Tonga. In modern terms, they were all in tectonically active subduction zones. No wonder they saw evidence of uplift; but, unknown to them, most of the atolls were on the aseismic, passive, subsiding plates of the Pacific and Indian Oceans.

Confirmation

The origin of atolls seemed perfectly clear in 1842, but even Darwin had not explained everything, so a century later no firm consensus was possible without drilling. Funafuti had been drilled to more than 300 meters, and the Japanese penetrated more than 400 meters on the island of Kita-Daito-Jima, but nothing but limestone was encountered, and the results were viewed as inconclusive regarding the origin of atolls. After World War II, the necessary doubly-rich patron for drilling appeared in the form of the United States Government. First, the Navy wanted to explode an atomic bomb in shallow water to see what happened to ships. That was done at Bikini Atoll in 1950. Next, the first thermonuclear device was exploded at Eniwetok (now Enewetak) in 1952. An enlightened scientific program was carried out in the Marshall Islands in connection with these tests. Drilling and geophysical mapping of the test atolls seemed particularly advisable. At Enewetak, the explosion was to be on an unprecedented scale and there was concern that the test island or even the whole atoll might slump into deep water. As we have seen in Chapter 5, enormous insular slumps certainly occur. The actual effect was to blast the test island into dust and produce marvelous sunsets for local oceanographers for some months.

Bikini was drilled to 775 m, encountering only reef and associated limestone in a normal stratigraphic sequence from middle Tertiary on upwards. Darwin's hypothesis was the only one compatible with the observations. Seismic refraction studies of the atoll showed that about 1300 m of coral rested on an irregular basement surface of material with the same seismic properties as the insular shelf of Hawaiian volcanoes. However, the basement material was not drilled, nor had more than 500 m of the base of the reef. Two holes at Enewetak were drilled to 2307 m and 2530 meters, where they sampled the irregular surface of basalt on which the atoll began to grow in Eocene time.

By 1965, Midway Atoll also had been drilled and the thickness of coral measured by explosion seismology at Enewetak, Funafuti, Kwajalein, Nukufetau, and Midway. The coral was 800 m to 1600 m thick at all the atolls but Midway, and in those that were drilled the top of the Miocene was only about 200 meters deep. The measured relief of the volcanic basement was 100 m to 600 m deep, and it was as irregular as the magnetic survey at Funafuti had suggested in the previous century.

The results of this great effort wholly confirmed Darwin's hypothesis. The coral was 800 m to 1400 m thick, was deposited in shallow water, and rested on submerged volcanoes with irregular, presumably eroded summits. Meanwhile, ever more proofs and corollaries appeared. For example, guyots were discovered in temperate waters, as Darwin and Dana predicted from their hypothesis of the origin of atolls. Now their argument could be inverted: if drowned ancient islands exist in cool waters, where are they in tropic waters unless they are covered by the coral of atolls? An elaboration of the basic hypothesis was also demonstrated in that erosional horizons were found by drilling. Thus the atolls did not always subside faster than sea level fluctuated. Indeed, the presence of fossils of land snails that live only on high islands (and uplifted atolls) suggested substantial relative uplift. Daly's fluctuating sea level had indeed played a major role in the Pleistocene history of atolls. But whether the earlier high islands were exposed to increased erosion by a major temporary drop in sea level and planed off by waves will not be known without further drilling. Meanwhile, it is possible that the apparent elevation was real and that individual atolls were raised by overriding midplate swells or by midplate flexing (described in the next chapter).

A curious misconception arose because of the discovery of guyots almost simultaneously with the drilling in the Marshall Islands. Indeed, Bikini atoll clearly rises from beside a guyot platform. It was assumed that the smooth, almost level, erosional surface of the guyot extended under and formed the base for the atoll structure. Thus, while stating that Darwin was confirmed, the geologists were substantially denying the whole logic of his synthesis. Their interpretation became widely accepted and even now a widespread impression exists that atolls rise from the wave-truncated platforms of guyots rather than, as Darwin said, from mountains eroded by rivers and protected from waves. This misconception should have been dispelled by subsequent drilling and seismic reflection studies, which showed an irregular basement under the coral, but it persists. Thus, it may be worthwhile to try to imagine some way that an atoll can form on a truncated volcanic platform. Clearly, the platform must be truncated in coral-free waters and then drift into or be invaded by a water mass populated by coral. The problem is that a wave-cut bank is almost at the maximum depth for lodgement of reef corals, and it continues to subside as it ages. Thus, to form an atoll on a guyot, it is necessary to elevate the guyot into very shallow water. This very probably occurs from time to time on midplate swells. Even now, some of the shallow guyots southwest of Tahiti may be moving up to where they can become coral banks before they sink again as growing atolls. Compared with Darwin's hypothesis, this seems to be a difficult way to produce atolls, but at least a few can indeed form on the antecedent summit platforms of guyots.

Evidence for subsidence of islands with barrier reefs

Island	Age (Ma)	Cliff height (m)	Mountain height (m)	Coral thickness (m)	Embayment	Outlying volcanic islands	Cliffs
Tahiti-Iti	0.4	300	1315		Filled bays	No	Widespread
Tahiti-Nui	0.5–1.2	60–150	2224		Filled bays	No	Widespread
Moorea	1.5–1.6	60–150	1207	280–340	Yes	No	North side
Huahine	2.0–2.6		669		Yes	Yes	
Raiatea	2.4–2.6	15	1035	320–340	Yes	Small	
Tahaa	2.6–2.9	15	590		Yes	Small	
Bora Bora	3.1–3.4		723	200–360	Yes	Yes	
Kosrae	4.0		625	160–300	Filled bays	One	
Mayotte	5.4		660	360	Yes	Yes	
Mangareva	5.2–7.2		425	680–1000	Yes	Many	
Ponape	8.0		786	600–800	Yes	Yes	
Truk	12.0–14.0		440	1060–1100		Many	

Erosion and Deposition

The erosional history of volcanic islands in warm tropical waters can readily be determined from examining the twelve with barrier reefs that have been dated and thus can be put in an age sequence (see the table on this page). The young volcanoes (0.4 Ma to 1.2 Ma old) have sea cliffs up to 300 m high and are gutted by headward erosion of streams and the formation of central calderas by erosion. Despite the rapid erosion, mountains remain that are as much as 2200 m high. There are no outlying small islands, because the submarine flanks are simple and steep. A reef circles much of the islands, but in some places it is fringing the shore and elsewhere separated from it by a lagoon. At the mouths of some of the principal rivers, erosional debris builds a delta on and kills the reef. Elsewhere, the rivers merely empty into the lagoon.

Above: A satellite image of Tahiti and Moorea (the smaller island). *Below:* Moorea's rugged topography, embayed river mouths, and fringing reef with the beginnings of a lagoon.

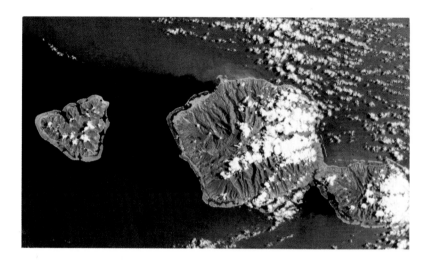

Moorea, west of Tahiti, is 1.5 Ma to 1.6 Ma old. The sea cliffs are 60 m to 150 m high although widespread, and the gutted interior has a peak 1200 m high. There are no outlying volcanic islands. The river mouths are embayed, indicating subsidence unmatched by deposition.

Five islands are 2.0 Ma to 4.0 Ma old. Only two have sea cliffs, and they are 15 m high. Most of the mountains are no higher than 600 m to

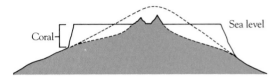

The outer slopes of a reef are typically much steeper than the slopes of the underlying volcano. The lower volcanic slopes may be extended upward to estimate the size and shape of the original island.

700 m. The valleys are almost all embayed, and there are outlying islands formed as small peaks become isolated from the main island.

There are four widely scattered barrier-reef islands 5.4 Ma to 14.0 Ma old. None have sea cliffs. The main mountains are 400 m to 800 m high; the valleys are embayed; and outlying islands are numerous. Indeed, the oldest, Truk, is sometimes called an "almost atoll," because it is a collection of small steepsided peaks in a vast lagoon.

All in all, we can hardly ask for more positive geomorphological evidence of the gradual subsidence of an initially cliffed and deeply eroded group of islands.

None of the barrier reefs have been drilled or surveyed with geophysical techniques. Thus, to estimate the rate of subsidence it is necessary again to appeal to geomorphology. The key is that reefs have steeper seaward slopes than the submarine slopes on which they lie. From topographic and bathymetric maps, the shape of the volcanic slope under the reef can usually be determined and the thickness of the reef then estimated. The method was used by J. D. Dana and W. M. Davis, mostly by extrapolating planeze slopes beneath the sea. Detailed charts of the sea floor near most reefs can now be used to extrapolate submarine slopes upward. This has been done for eight of the twelve dated islands with barrier reefs. The thickness of coral estimated in this way increases from between 280 m and 340 m for an island with an age of 1.5 Ma to 1.6 Ma, to between 600 m and 1000 m for two islands aged 5.2 and 8.0 Ma, to about 1100 m for 14-Ma Truk. The data thus are consistent with gradual subsidence. It should be noted that very similar thicknesses would result if the reefs merely grew sideways into deeper water without any subsidence. Lateral growth certainly takes place on some reefs, as Darwin showed, but it seems unlikely that it is important. Among other reasons, the islands apparently subside roughly 300 m to eliminate the sea cliffs and hundreds of meters to embay the valleys and isolate the small peaks.

If we accept the reef thickness as a measure of subsidence, the rates of subsidence of the islands can be compared with those expected from cooling of the lithosphere. The islands subside as fast as if they had been built on lithosphere with a thermal age of no more than 3 Ma, despite the fact that the actual age of the underlying lithosphere is early Cenozoic or Mesozoic. Some of the subsidence might be attributed to the added mass of the reefs, but the reef structure is very open and the aggregate density is too small to have much effect. Thus, most of the subsidence apparently is thermal and the rates indicate that the lithosphere has been thermally rejuvenated on midplate swells.

The subsidence of islands with coral appears to be entirely different from the subsidence of those without. Indeed, it would be very surprising if they were the same. A volcanic island in cool water is eroded away by weathering, wave, and river, and almost all the erosion products are carried away from even the submarine base of the volcano. Thus, isostasy keeps the island at sea level until it is truncated. Then it sinks with the cooling sea floor, but by that time even initially very young crust has aged 5 Ma to 10 Ma and is not sinking so fast. In contrast, a reef-girt island is protected from waves, and the widening lagoon captures erosion products that are not in solution. Thus, isostasy does not keep the island at sea level, and it sinks immediately with the sea floor, which may be very young thermally.

Drowned Atolls and Banks

When he was on the *Beagle,* Darwin considered himself primarily a geologist, but, of course, he thought as a naturalist and always had the competition and the death of organisms in the back of his mind. He observed that submerged banks with flat tops and raised edges were dispersed among atolls in some regions. A notable one is Great Chagos Bank, in the central Indian Ocean. The bank is roughly 150 km in diameter. The interior is a "level muddy flat" somewhat less than 100 m deep, surrounded by long submerged banks, about 30 m deep, made of coral sand but very little live coral. The deep banks in turn are surmounted by a series of long narrow banks, at a depth of 10 m to 20 m, which form the rim of the whole great feature. This rim is composed of dead reef limestone with a thin layer of sand but scarcely any live coral. In short, Great Chagos Bank appears to be a dead and submerged atoll, despite the presence of nearby atolls and even of some thriving coral pinnacles within the great drowned lagoon. Why should a reef not die? Darwin wrote, "it cannot be expected that during the round of change to which earth, air, and water are exposed, the reef-building polypifers should keep alive for perpetuity in any one place." Particularly if oceanic islands subside.

Since Darwin's time, scientists have become more specialized and geologists generally have had difficulty in understanding how atolls could be drowned. The problem is not trivial. There are 261 atolls in deep ocean basins and scattered among them are 116 banks, 10 m to 20 m deep, that are suitable sites for atolls and many of which have the morphology of drowned atolls. Moreover, many atolls, particularly in a few regions, are incomplete rings, and some are only single small islands on the edge of extensive banks. In sum, almost a third of the potential sites for atolls are unoccupied, so it appears that atoll mortality has been high in recent geological time.

The cause of the high mortality may be related to the one important fact about atolls that Darwin did not know. Only 12,000 years ago, atolls were all islands 100 m to 150 m high because sea level was low. At that time, all the exposed coral was dead but, presumably, patches of live coral formed a fringe around the shoreline of most islands. Then sea level began

Darwin's chart and cross section of Great Chagos Bank.

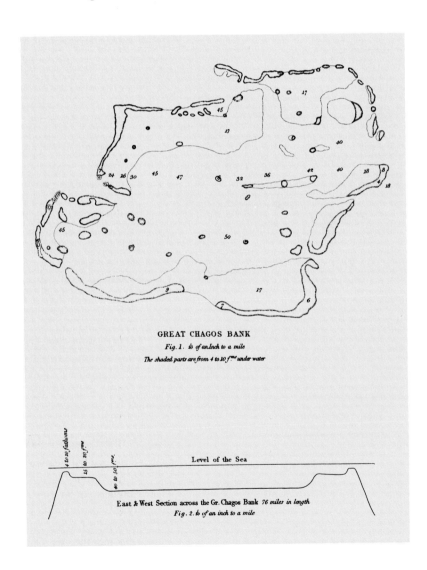

GREAT CHAGOS BANK

Fig. 1. ¹⁄₁₀ of an Inch to a mile

The shaded parts are from 4 to 10 f ᵐˢ under water

Level of the Sea

East & West Section across the Gr. Chagos Bank *76 miles in length*

Fig. 2. ¹⁄₁₀ of an inch to a mile

to rise at rates of 6 mm/yr to 10 mm/yr. Darwin had showed that small patches of coral in optimum conditions can grow 360 mm/yr, but he lacked data for whole reefs. Estimates by Harry Ladd in 1961 put the overall rate of reef growth at less than 14 mm/yr. Even so, this rate could have kept up with rising sea levels. Nonetheless, about a third of oceanic reefs did not, or have not yet, caught up with sea level, so conditions must have been less than optimum for reef growth. Such conditions are observed in the passes through the reefs of thriving modern atolls. In general, the passes do not fill in because of the relatively variable salinity, temperature, and sediment load of the waters that flow in and out in tidal cycles.

The factors that controlled reef growth are uncertain, particularly where atolls and shallow banks are intermixed. However, most drowned atolls are in regions that are relatively free of healthy ones. Such regions are in the western Caroline Islands, the South China Sea, and the Melanesian region north and west of Fiji. In the last of these regions, some drowned atolls have been dredged and their substructure determined by explosion seismology. They are thick reefs that had been atolls for tens of millions of years, but now there is no sign of them at the surface. Except for a scattering of non-reef-forming corals, the drowned atolls are dead. Indeed, no live coral pinnacles such as exist on Great Chagos Bank rise from the great flat "lagoons." It is perfectly safe to proceed at cruising speed over seemingly endless rocky bottom only 20 m below the hull. However, if the oceanographer loses his nerve in the middle of one of these banks, it can take half a day at prudent speeds for maneuvering over a reef to finally reach deep water.

One modern characteristic of the regions with drowned atolls is that the rainfall is unusually high. Tropical rainfall is highly variable with latitude, and atolls experience rainfall ranging from 1000 mm/yr to 5000 mm/yr. However, most atolls are exposed to less than 2500 mm of rain per year. In contrast, drowned atolls are concentrated in regions with 2500 mm/yr to 4000 mm/yr of rain, so low salinity may be one of the factors that prevent drowned atolls from growing to the surface.

Coral-Capped Guyots

Thermal subsidence is constantly pulling the drowned atolls downward and presumably some of them will eventually become submerged deep enough to be guyots. Certainly, as Darwin surmised, there are dead, deeply drowned atolls. The first of these discovered in the central Pacific was appropriately named Darwin Guyot in 1974. It has a fossil reef fauna, the morphology of an atoll, and a minimum depth of 1250 m. Other guyots that are presumably drowned atolls have not been sampled. However, seismic reflection profiles across some guyots in and north of the tropical Pacific show mountainous basement topography buried by more or less layered rock with a relatively flat upper surface. This rock can hardly be anything but a coral reef that was deposited over a subsiding volcanic island. Thus, Darwin is wholly confirmed. Atolls may die and sink beneath the waves.

8

Midplate Interactions

Islands have a strong tendency to cluster, and the hot-spot model for generating oceanic volcanoes readily explains many of the facts about clusters of islands. The Canary Island hot spot, for example, lies near the pole of rotation of the relatively motionless African plate. The hot spot produces volcanoes from time to time, and inevitably they are in a cluster. Hot spots also produce island clusters on the rapidly moving Pacific plate, but for a different reason. If the ocean were removed, the products of the Hawaiian hot spot would be seen as an enormously long line of volcanoes that disappears into the Kuril-Kamchatka Trench. However, erosion and subsidence have caused almost all of this line to be concealed by the waves, and thus the young, high islands appear as a relatively short and wide cluster. However, a simple, single hot-spot model cannot explain one of the most striking features of clusters, namely, that modern and ancient islands of highly divergent ages are commonly mixed together.

INTERMIXING OF ISLAND TYPES AND AGES

Isotopic and fossil dating, as well as morphology, give evidence that the islands of many clusters are mixed in age. With regard to dating, the high Austral Islands, for example, show only weak evidence of an age progression in the direction of plate motion. Instead, it appears that, for roughly the last 20 Ma, active volcanoes have erupted there more or less at random along a line. The mixture of ages in the Hawaiian region appears to have an entirely different cause. Three ancient hot spots produced parallel lines

The cliffs of Makatea, an elevated atoll in the Tuamotus, are an ancient upraised coral reef. In the foreground is a more recently emerged reef flat.

of volcanoes on the Pacific plate about 70 Ma to 80 Ma ago. Apparently none of the volcanoes was ever high enough to be an island. These three lines of volcanoes have drifted with the plate (which has since changed direction), and two of them have crossed over the Hawaiian hot spot. As a consequence, the ancient volcanoes are mixed among, and doubtless in places buried by, the voluminous volcanoes of the Hawaiian chain that are tens of millions of years younger.

Mixed ages in a cluster can also be seen by comparing the youth of high islands with the presumably greater ages of nearby atolls and guyots. Pitcairn Island, for example, is a high, youthful (although extinct), volcanic island in the midst of the numerous atolls of the southern Tuamotus. Likewise, Kusaie, Ponape, and Truk are relatively youthful high islands with barrier reefs among the Caroline atolls. The Austral Islands show a similar pattern. Rapa is a high, deeply eroded volcano about 5 Ma old. It rises above a group of enormously broad guyots which have been submerged 1300 m to 1600 m and probably are more than 40 Ma old.

The significance of the morphological line of evidence was appreciated almost as soon as guyots were discovered. In 1944–1945, Harry Hess found guyots mixed in among the atolls of the Marshall Islands. Having deduced from their flat tops that guyots are ancient islands, he was immediately required to explain why the guyots had not become atolls when they subsided. There were two simple explanations, and for present purposes the important point is that both of them required that the atolls and guyots in the cluster had to have very different ages. Either the guyots were so old that reef corals had not yet evolved when the islands began to subside, or else the islands had existed in temperate waters and had subsided too deep for lodgement of reef corals before the islands drifted into the tropics. Continental drift was in disfavor when Hess published, in 1946, so he deduced that the guyots were probably at least 500 Ma old. Only fourteen years later, he conceived of the idea of sea-floor spreading—in part to explain how guyots might subside and also drift about.

Guyots and atolls also are mixed together in the Austral, Caroline, Gilbert, Line, Phoenix, and Tuamotu island clusters. In short, on the vast, rapidly drifting Pacific plate, almost all clusters of modern and ancient islands have mixed ages that indicate more than one period of voluminous, shield-building volcanism. This circumstance is in apparent contrast to other oceanic plates. The only significant mixing of ages in the Atlantic, in the Canary Islands, has already been explained by the immobility of the African plate. The Indian Ocean lies over three tectonic plates; two of these have been relatively fixed, but the Indian plate has drifted very rapidly. All types of modern and ancient islands abound in the Indian Ocean, but to a remarkable degree the types do not mix. Two great lines of ancient volcanoes mark north-south hot-spot tracks across the Indian plate and confirm the northerly drift demonstrated by paleomagnetism. However, the western line is all atolls and the parallel eastern line is all guyots. The only mixing of ages of any consequence is northwest of Madagascar, where active and extinct volcanic islands lie among atolls.

Causes of Mixed Ages

The ages of islands in a cluster are mixed because shield-building, voluminous volcanism occurred at intervals separated by tens of millions of years. This can be seen to be happening now in the western Pacific and Polynesia. Several island-generating hot spots exist near the crest of the East Pacific Rise, on the eastern edge of the great Pacific plate. As the plate and islands drift northwest, they sink until they enter what may be called the "Polynesian plume province." The plate is elevated here and there as it overrides the midplate swells of the province, and active volcanoes grow among the atolls that themselves grew near the crest of the East Pacific Rise. Judging by the present geology, the reason that there are mixed ages in island clusters on the Pacific plate is simply that hot spots are distributed widely under the plate, which is big enough that a cluster that originated over one hot spot has a reasonable chance of drifting over another.

It might seem that the only question that arises from the observation of mixed ages is why there are so many hot spots in the region—a question that is fortunately beyond the scope of this book, because no one knows the answer. However, there are two interesting questions within our scope that arise not from what is observed but from what is not observed. Why are there gaps between clusters? Why, if the oceanic crust is elevated a kilometer or more while overriding a midplate swell, do we not observe groups of atolls elevated a similar amount above the sea?

With regard to the first question, a thought experiment may be useful. Consider a large piece of plasterboard lying on a frictionless surface. Let it make a random walk under a randomly placed array of spray guns that spray small circles of paint now and then. There is always a possibility that one circle will be superimposed on another. However, as time passes, the chance that all the circles would be superimposed and large unsprayed areas would persist becomes miniscule. Does this thought experiment differ in some important factor from the generation of island clusters in the Pacific? It might be questioned whether the plate motion is random, but it has changed direction at least four times. Perhaps the hot spots are not randomly distributed, but, if not, no one has yet identified a pattern. Nor is it reasonable that insufficient time has elapsed to fill in the gaps. Many of the mixed clusters were first active 60 Ma to 100 Ma ago.

Shallow-water fossils dredged by E. L. Hamilton from the tops of guyots in the western central Pacific.

It appears that the thought experiment can approximate nature only by adding sensors to the spray guns so that, after the first time, they ordinarily turn on only when they are over older circles of paint. Consider now the vulnerability of the Pacific plate to volcanism. The regions where volcanic clusters initially develop tend to be on thermally rejuvenated midplate swells. Thus, a cluster may be more vulnerable to further volcanism than the surrounding region even after the cluster drifts away from the initial hot spot. Intense hot spots may break through anywhere; however, less voluminous hot spots may be able to penetrate the lithosphere only where it is relatively vulnerable. In these circumstances, volcanism would occur repeatedly in island clusters but rarely between clusters, just as is observed in the western Pacific.

Let us turn now to the absence of groups of highly uplifted atolls on midplate swells. The most conspicuous midplate swells in the Pacific underlie the Hawaiian, Marquesan, Samoan, Austral, and Society islands. As it happens, there are no older atolls anywhere near the Marquesan or Hawaiian swells. On the other hand, seamounts on the Hawaiian swell are notably shallower than those in the surrounding region, and this might be a consequence of uplift. There are no atolls uplifted by overriding the Austral or Society swells, either. However, there are guyots that appear to be uplifted, and the amount and distribution of uplift may indicate why highly uplifted atolls are absent from swells.

The islands that grow near the crest of the East Pacific Rise are in cool, non-coralliferous water and so become guyots that sink with the lithosphere once they are truncated by erosion. By the time these guyots drift to the southeastern edge of the Society swell, their summit platforms are roughly 1300 m deep. Another group of guyots lies on the western flank of the swell, where it appears that the sea floor has been uplifted by roughly 600 m, although the abundance of guyots confuses the bathymetry. These guyots have been surveyed with seismic-reflection techniques, and they appear to be deeply drowned atolls, mountainous volcanoes buried under flat-lying coral reefs. The reefs in turn are buried under thin pelagic ooze. The summits of the ancient atolls, under the ooze, are at depths of 600 m to 1000 m. Thus, compared with the guyots just southeast of the swell, they appear to have been uplifted 300 m to 700 m, or about the same as the sea floor. Similar, but less documented uplift of guyots apparently occurred on the Austral swell. In sum, judging by present geology, most guyots, including deeply drowned atolls, are so deep by the time they drift over a midplate swell that 1000 m of uplift may not bring them up to the surface.

Is present geography and geology representative of the past? Consider the guyots of the western central Pacific. Dredging, largely by E. L. Hamilton and Bruce Heezen, showed that fourteen of the guyots were islands simultaneously, or nearly so, about 100 Ma ago. The lithosphere was 13 Ma to 64 Ma old under these coral-clad islands when they died and began to subside at the same time for some unknown reason (a meteorite splash?). Young lithosphere subsides more than old, so if the guyots merely moved down with a cooling plate their summits should now be at depths ranging roughly from 300 m to 2000 m. In fact, the depths are remarkably concordant, in the range of 1000 m to 1300 m. This can be readily explained if the guyots of various ages were formed as the ancient plate drifted over a series of midplate swells, and the final one thermally rejuvenated the lithosphere under all of them to 50 Ma.

ELEVATION AND SUBSIDENCE

There are, in fact, elevated atolls in Polynesia, but their distribution among normal atolls shows that they are not elevated because they override midplate swells. Early European explorers noted a few score of flat-topped islands 10 m to 100 m high and ringed by cliffs. They look very much like cakes that failed to rise properly. However, this type was rare compared with mountainous volcanic islands and atolls with scarcely any relief. Lyell remarked on the unusual character of these islands even before Darwin put to sea on the *Beagle*. Darwin himself sailed near one of the most interesting of these islands, which was then called Metia or Aurora and is now called Makatea. (This annoying practice of changing names

View of Makatea from the northwest, showing the cliffline and the upper plateau covered with dense vegetation.

Notches in Makatea's cliffs.

has accelerated recently as new nations arise in the Pacific.) However, his published journals make no mention of it. He does report for 15 November 1835 that the distant peaks of Tahiti were in sight at dawn. Presumably, he passed Makatea during the night.

A few years later, James Dwight Dana became the first geologist to land on Makatea. The island was an uplifted coral atoll with prominent wave-cut notches cut into the encircling cliffs. The whole resembles an amateurish cake with three layers. Dana personally measured the height of the cliff, and this became important when the cause of the uplift was realized, more than a century later.

Flexing of Intermixed Islands

Darwin was the first to propose a cause for the uplift of many coral islands. He observed on his global map of different types of coral reefs that many uplifted reefs lay in or near regions of active volcanism. He proposed that such regions generally were elevated by expansion in the earth's interior, just as other regions, with atolls, subsided by internal contraction. In modern terms, he was discussing the elevation of such islands as Guam in tectonically active subduction zones. He offered no explanation for the few uplifted atolls within the subsiding region of Polynesia.

Such an explanation was given by L. J. Chubb as a consequence of his observations on the *St. George* Expedition, which took him to the Austral

The cliffs of Rurutu.

Islands in 1925. He found that Rurutu and Rimatara both had fossil fringing reefs that were uplifted 150 m and 8 m respectively. He observed that other, unvisited islands nearby lay on a line with Rimatara and appeared similar to it on maps. Another parallel line of higher but extinct volcanic islands lay not far to the south. Chubb proposed that a pair of large crustal folds or flexures of the oceanic crust were drifting to the northeast. First the trough passed under the older volcanoes of the region, causing subsidence and reef development on the islands. Then the swell came through and elevated the reefs. For some reason, the younger volcanic islands occurred on the southern, subsiding, limb of the swell. This was quite a satisfactory explanation for the scanty data available and one of the few ideas that sought significance in the fact that only a few coral reefs in Polynesia were elevated.

In the succeeding decades, all elevated coral reefs were identified, and many were mapped topographically and even geologically. The accumulated evidence came to be interpreted in terms of eustatic sea levels. The present high levels of coral were taken to be caused by past high levels of the sea. The fact that almost all atolls showed no evidence of significantly higher levels of the sea was deemed inconsequential.

There the matter stood when new, apparently unrelated kinds of information began to accumulate. Several active submarine volcanoes were detected in Polynesia by remote sensing, and all the high volcanic islands in Polynesia were dated by the potassium-argon method. In 1976, I had

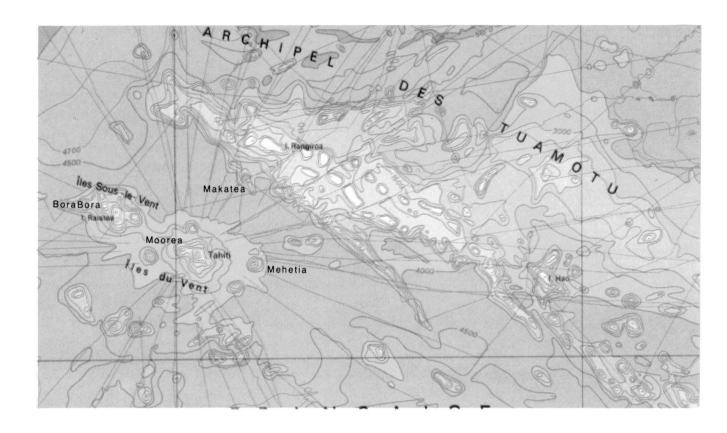

occasion to prepare a series of large-scale charts of Polynesia that identified all different types of seamounts and islands, including their ages, if known, and the depths of guyots and elevation of coral reefs. It was obvious that all the elevated reefs were near active volcanoes or high islands less than about 1 Ma old. But not all atolls near young volcanoes were elevated, so the cause could not be a regional bulge due to volcanic heating or over-riding a midplate swell.

Except for heating and cooling, the only demonstrated way to change the depth of the oceanic lithosphere is by loading and flexing it. If the moat and arch around Hawaii are taken as a model, both the uplift and lack of uplift of atolls in Polynesia can be explained. The atoll near young Pitcairn is not uplifted because it is in the moat. More distant Henderson Island is uplifted because it is on the arch. And so were explained the high uplift of Makatea and the low uplift of certain atolls in the Tuamotus by young Tahiti and Mehetia. Likewise, the high uplift of Mangaia and the low uplift of several of the southern Cook islands by young Raratonga. Uplift just depends on where an island is on the circular arch produced by a young point load. Of course, the same physical effects followed the same cause in the past, but, given time, all islands tend to sink. Thus it is that the numerous atolls around the Gambier Islands show no sign of uplift. The Gambiers are a group of volcanic islands within a large barrier reef. The main volcanic platform was active 4.4 Ma to 6.0 Ma ago. Any atolls elevated around the point load apparently have had time to return to sea level because of the continuing subsidence of the lithosphere.

It is one thing to determine a cause and quite another to quantify and generalize it. Marcia McNutt analyzed these observations and calculated the flexural rigidity of the lithosphere that would cause the observed uplift. The fit of theory to observation was remarkably good. Theory gave an uplift of 9.5 m for Rimatara, and 8.0 m was observed; the predicted uplift of Makatea was 71.9 m, and 70.0 m was observed; and so on. It was hard to believe that uplifted coral islands would respond so uniformly to erosion or even be immune to it, and indeed, some of the quantitative aspects of our joint paper have been questioned because the basic geographic observations leave much to be desired.

The problems in observation are several. First, the uplifted islands are remote and lack harbors; most of them even lack airfields and are rarely visited by commercial transport. It has taken me 35 years of oceanographic exploration on Scripps ships to land on or survey around a majority of these uplifted atolls.

Second, the records of the heights of islands are generally rather casual. Naturally, mariners and scientists prefer to make accurate measurements, but it is not easy. The islands are ringed with cliffs, and normal land surveying methods are difficult to use. Cliff heights, in fact, can be measured rather accurately from a ship, provided either the distance from shore or the height of some object on shore is known. However, the height of the cliff may vary, and the interior of the island may contain broad elevations or depressions. Which of the topographic features should be accepted as indicating the uplift of the island?

The History of a Measurement

Charles Wilkes was leader of the U.S. Exploring Expedition of 1838–1842; in his five-volume narrative about it appears the following about Makatea Island: "On approaching its eastern end, I sounded at about one hundred and fifty feet from its perpendicular cliff, and found no bottom with one hundred and fifty fathoms of line. The cliff appeared worn into caverns. We landed close in its neighbourhood, and on measuring its height, it proved to be two hundred and fifty feet." In modern units, something was 76 m high, but the pronouns permit it to be either the cliff or the island. Dana, who made the measurement, shortly clarified the point. It was the island.

In those days, only scientists who were or would be famous had the luck or resources to reach remote islands on exploratory expeditions. The next one to appear was Alexander Agassiz (son of the great Louis), who had deliberately spent his early career making money as manager of a copper mine in Michigan so he could retire in mid-career to do private research. However, he had connections, and he had the U.S. Fish and Wildlife ship *Albatross* under his direction when he approached Makatea, at the turn of the century. He landed and took samples and photographs all around the island. He said that it was 70 m high and that the cliffs ranged from 36 m to 61 m high.

William Morris Davis, the greatest geomorphologist of his time, published a whole book on coral reefs in 1928 and in it gave the height of Makatea as 70 m. The *U.S. Navy Sailing Directions*, the ultimate, official American governmental source on the subject, said the height was 70 m in the edition on board our Scripps ship when we surveyed around Makatea in 1952. Thus, I had had no doubts about the height of the island for some decades when it accidentally became important in estimating the flexural rigidity of the lithosphere, in 1976.

Meanwhile a far better but, regrettably, contradictory data set was collected for commercial reasons. Most of the elevated coral islands contain large rich deposits of phosphorite ores. These were already being mined very early in the present century, and mining required the construction of buildings, water works, narrow-gauge railroads, and elaborate loading docks suspended over the deep water at the base of the cliffs. Such construction in turn required detailed topographic surveys, and so a sketchy topographic map of Makatea, without units, appeared in a German report on insular phosphate deposits in 1913. La Compagnie Francaise des Phosphates de l'Oceanie made a more informative map available in 1934. There were three peaks in the interior with heights of 79 m, 93 m, and 113 m. The cliffs were generally 70 m high. Apparently Dana, Wilkes, and Agassiz were careless in making or interpreting their field notes.

The topography, geology, and climate of Makatea were accurately described for the British Naval Intelligence Division in one of a series of reports in 1943. Elevations of peaks, the central plateau, and terraces notched in the cliffs were all included. As it happened, I was in American intelligence and working with British naval intelligence at that time, and I read the reports. However, one cannot note everything, and the reports were classified. Thus it was that the militarily unimportant height of an obscure island far from the combat zone had not been corrected in the *U.S. Navy Sailing Directions* by 1952.

In 1981, Kurt Lambeck introduced the correct topographic data for Makatea into his analysis of the flexing of the lithosphere. However, he obtains the best fit with data from other islands if the uplift of Makatea is only about 45 m. In fact, this is about the height of some of the conspicuous notches cut in the cliffs, but then the additional height of the island is a major anomaly. However, in the McNutt and Menard analysis, the height of Mangaia, in the Cook Islands, is also a major anomaly. It is suspected that the load of undiscovered young seamounts may be complicating matters, but, fortunately, the two analyses are rather consistent regarding the stiffness and thickness of the elastic upper layer of the lithosphere. The rigid, drifting lithosphere in the western Pacific is 70 km to 100 km thick, but the elastic layer that flexes in response to loading is only 10 km to 14 km thick, on average, in Polynesia and Hawaii. At the limit of present knowledge, it appears that flexural rigidity varies on a midplate swell in a manner that can be explained by heating and thinning as the lithosphere moves over a hot spot. But the application of island history to the study of physical properties of the lithosphere has barely begun.

Elevation and Subsidence While Sea Level Fluctuates

The geological history of subsidence and elevation of high coral islands is notoriously difficult to interpret. Even the paleontological age of the uplifted coral is questionable for most of them. Islands do not stay elevated for very long, by geological standards, if they are on midplate swells. The coral of which they are made ordinarily accumulated because of subsidence during an interval that ended with uplift. Consequently, the coral itself may be only 1 Ma to 20 Ma old, or of Miocene or Pliocene age in the geological time scale. The fossils preserved from these epochs on the Pacific islands tend to be less than ideal for assigning paleontological ages. Moreover, the surface outcrops on uplifted islands are exposed to percolating water, and fossils commonly are dissolved.

We may look to Makatea Island again for an example of the history of paleontological dating of an uplifted island. Dana reported that it was made of solid limestone without diagnostic fossils and hardly any trace of corals. Agassiz and the paleontologist Dall had examined fossils from the elevated reefs of Fiji, and, "from the evidence of the fossils and the character of the rock," they regarded the reefs as of Tertiary age. Concerning Makatea, Agassiz ventured paleontological dating by remote sensing: "From the very description given by [Dana] of the character of the cliffs and of the surface of Makatea, I felt satisfied that it was composed of the same elevated coralliferous limestone so characteristic" of Fiji and which "Mr. Dall and myself have been led to regard as of Tertiary age." The copy of Agassiz's monograph of 1903 in the Scripps library originally belonged to the distinguished expert on corals and erstwhile director of Scripps, T. Wayland Vaughan. Presumably, it was his hand that noted "no evidence" in the margin next to the above quote. Agassiz wrote that he and Dall collected fossils on Makatea but Dall apparently never described them. Moreover, the fossils are not in the collections of Harvard's Museum of Comparative Zoology, which was established by Louis Agassiz.

Present-day descriptions of the island by French geologists do not mention Tertiary fossils, but they do report modern (Quaternary) molluscs in the limestone at the top of the cliffs. That does not prove that the rock at the base of the cliffs is not Pliocene or Miocene. It does prove that the island was elevated in the last million years or so—which agrees with the hypothesis that the cause was lithospheric flexing by the load of young volcanoes.

If all the high islands in Polynesia were uplifted in the past million years, they should show the effects of the eustatic fluctuations in sea level during that period. Such effects on rising islands in tropical subduction zones were described in Chapter 4, "Sea Level." If the island is being elevated at a constant but slow rate, a rapid eustatic rise in sea level results in the growth of fringing reef. A drop in sea level kills the reef, which continues to be elevated above sea level. The assemblage of dead reefs due to successive eustatic fluctuations resembles a staircase superimposed on the relatively smooth and gentle slope of the island.

Somewhat similar phenomena are reported from the uplifted Polynesian islands if due account is taken of their steep slopes. Rurutu Island, in the Austral group, was active over the Macdonald hot spot about 12 Ma ago. Then it was eroded almost to sea level except for a peak about 150 m high. Blocks of limestone on the broad summit plateau, which is now about 240 m high, confirm that most of the island was once submerged. However, the most conspicuous feature of the island is an elevated fringing reef that forms a coastal barrier with almost vertical cliffs 60 m to 90 m high. Indistinct bedding in the reef is horizontal, and the reef rests unconformably on a conglomerate of basalt cobbles, which in turn lies on relatively unweathered basalt bedrock. It appears that Rurutu was deeply eroded in waters too cool for vigorous reef growth. Then the island was elevated relatively rapidly when sea level was constant or falling. This

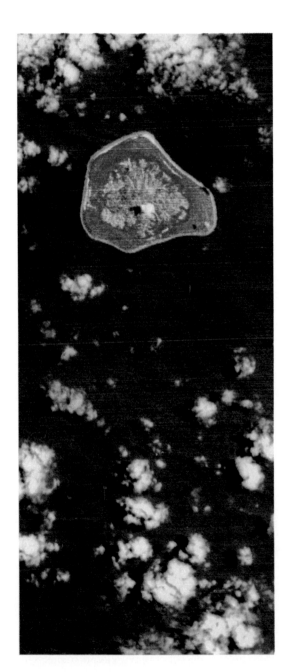

The raised atoll of Mangaia by satellite photography.

coated the former insular shelf with a thin layer of limestone that has since been eroded into isolated blocks. During the next period, the island presumably continued to rise, as it did before and after. However, sea level rose relatively rapidly and a fringing reef developed that was 60 m to 90 m thick. Then sea level fell and the island rose, and a single great steplike reef was exposed.

Mangaia Island, in the Cook group, has had a more complex history than Rurutu since it, too, originated over the Macdonald hot spot, 17 Ma to 18 Ma ago. The topography of the island is extraordinary. The deeply weathered hills of the central interior have gently sloping or nearly flat tops with a distinct common level of about 167 m. This central region, about 5 km in diameter, amounts to a moderately eroded hilly plateau and is surrounded by a ring of swampy depressions that are only a few meters above sea level. However, they are separated from the sea by a great circular wall of limestone 55 m to 70 m high and 500 m to 1000 m wide. The wall, which is an uplifted barrier reef, looks for all the world like a giant structure built to keep sea monsters out, just as in the motion picture a giant wall kept out King Kong. Access to the fertile interior is along a short, narrow road in a sloping artificial trench. A village lies at the outer end of the road, on a slightly elevated marine bench. Similar although less elevated limestone walls are found elsewhere in the Cook Islands, and each is called a *makatea*, which is confusingly the same name as the elevated island in the Tuamotus. The outer side of the makatea of Mangaia has relatively horizontal benches and notches at heights of 3 m, 5 m, 14 m, 23 m, 36 m, and 56 m. The notches were produced by solution of the limestone at sea level, and the benches by upward and outward reef growth that formed steep steps on a pre-existing mid-Tertiary reef.

The indicated, although still uncertain geological history perhaps began with a volcanic island that was eroded to sea level while in waters too cool for reef growth. It was then that the future central plateau was leveled. Next, Mangaia was elevated in warmer waters; a fringing reef formed; the island began to sink again, and the fringing reef was transformed into a barrier reef encircling a lagoon around a very low volcanic island. Then, from about 2 Ma to 1 Ma ago, the growth of volcanic Raratonga elevated the island. However, the load of Raratonga is not adequate to account for all the uplift, so another, wholly submarine volcano is required. Such a seamount exists at the right distance, but its age is unknown, so the uplift may have continued to the present. Corals in the position of growth exist in reef rocks almost 2 m above sea level. Herbert Veeh has determined their ages isotopically as about 100,000 years. The observations suggest continuing recent elevation. At the same rate, the emergence of the whole barrier reef would have taken roughly 3 Ma.

While the island was slowly rising, sea level repeatedly rose and fell more rapidly in Pleistocene time. However, the outer slope of the Tertiary reef was too steep for the formation of a succession of reefs like a staircase with broad treads. Nonetheless, the series of notches and benches is caused by the same phenomena that produce reef staircases elsewhere.

The island of Makatea was also elevated while sea level fluctuated, but the resulting morphology is different from the other high islands in Polynesia. The encircling cliff is almost vertical, so successive high stands of sea level could only cut the notches that are so evident in the cliff face. The origin of the notches poses no difficulties, but the origin of the cliff is another matter. William Morris Davis concluded that the vertical cliff, like all other such sea cliffs, was a consequence of wave erosion. Moreover, he observed that the whole cliff had to exist before it was possible to start cutting notches. Thus, Makatea apparently had a complicated history that began with subsidence of an inactive volcano in coral seas and formation of an atoll. Next, the atoll was elevated to about the present height, and the 30°–40° slopes typical of an atoll were eroded into a vertical sea cliff. It is possible that this elevation occurred on an arch produced by the load of the northern group of Society Islands, which include Bora Bora and Raiatea. If so, the uplift was 3 Ma to 5 Ma ago and the lithosphere was stiffer than it is now. Next, the island subsided rapidly without coating the old reef with a layer of young reef. Finally, the island was gradually elevated to its present height, as the cliff was notched.

This history seems very complicated, and it is not accepted by some geologists in Polynesia. However, other lines of evidence also indicate repeated submergence and emergence of some high islands. Elevated limestone atolls are exposed to solution by rainwater, which generates a typical kind of small-scale erosion surface called a *karren field*. This consists of fantastic pinnacles of undissolved limestone, 2 m to 5 m high, which on many islands are buried in "terra-rosa" clay, presumably the residuum from solution of higher layers of limestone. This typical red clay is not found on Makatea, Ocean, Nauru, or Christmas islands, which are the best known high coral islands because they have all been mined for phosphate. G. Evelyn Hutchinson, who made a monographic study of phosphate deposits on islands, proposed that the red clay was once part of the karren field on these islands but was washed away when the islands were submerged before their present emergence and elevation. The cycles of submergence, emergence, resubmergence, and present reemergence are beyond dispute with regard to Ocean and Nauru islands. Mining has exposed the old karren fields because the limestone pinnacles do not contain phosphate ore. The islands were emerged when the karren fields developed, but coral heads are

A karren field on the upper plateau of Makatea.

attached to some of the solution pinnacles on Ocean Island. Similarly, shorelines have been cut into pinnacles on Nauru Island. Thus, the islands were submerged between two periods of emergence.

The evidence from Ocean and Nauru seems to support the hypothesis that Makatea has twice gone down and up. However, Ocean and Nauru are the only two high coral islands whose cause of elevation remains unexplained. Perhaps the two islands bob up and down because of tectonics. On the other hand, perhaps the last submergence merely was the result of a rapid rise in sea level overtaking the continuing rise of the islands.

Phosphorite Ore on Uplifted Coral Islands

Isolated uplifted coral islands with vertical sea cliffs and iron-bound coasts are inhospitable for almost all species. Thus, they make ideal rookeries for sea birds; from that circumstance, many have become commercially important sources of phosphorus, in the form of either guano or the phosphorite rock that evolves from it under some conditions. Small deposits of fossil guano were once mined from some of the slightly elevated atolls of the central Pacific, but, by the end of the nineteenth century, most of it was gone.

It was Sir John Murray who first realized the potential of the high islands that have been major world sources of phosphate for the past eighty years. Murray was brought up on the plains of central Canada but traveled to Scotland, his father's homeland, to study when he was seventeen. He never obtained a degree, but at age 31 he sufficiently impressed Sir Wyville Thomson, the organizer of the *Challenger* Expedition, to obtain a position as junior scientist. He spent much of the time from late 1872 to 1876 at sea, and by default he was made responsible for the collection and analysis of deep sea sediments. Allowing for inflation, the *Challenger* was probably the most expensive oceanographic expedition that ever sailed. After its return, the British Treasury allotted funds for analysis and publication of results, and Murray was part of the small permanent staff. He became leader of the project when Thomson died. Volume after volume of great grey-green monographs poured out, but the Treasury stopped its funding in 1889, even though much remained to be done. At age 48, John Murray was unemployed.

In that year, Murray married Isabel Henderson, the only daughter of the owner of the Anchor Line, operating steamships out of Glasgow. From all accounts, Murray was a devoted husband and father, and his wife proved fortunate in yet another way that is rare for heiresses. He immediately became an imaginative, patient, diligent, and brilliantly successful entrepreneur. One of Murray's shipmates from the *Challenger* happened to be on H.M.S. *Egeria* in 1887 and was a member of the shore party that landed on uninhabited Christmas Island in the Indian Ocean. (There was another island that had the same name in the central Pacific, but its name has been changed to Kritimati.) He sent a small rock sample from the island to Murray, who did a chemical analysis. It was a very rich ore of phosphate.

Murray immediately realized the implications of his find, and, in the same year, he persuaded the British Government to annex the island. It was 300 kilometers southwest of Java, isolated, and not of the slightest interest to anyone else. Four years later, Murray and a Mr. Koss of the Cocos Islands obtained a lease of the island. At his own expense, Murray sent C. W. Andrews, of the British Museum, to survey the island in 1897–1898. Construction of a railroad and docks followed, and exploitation began in earnest about 1900. The results of this investment were dazzling. When Sir John Murray, K.C.B., was killed in an automobile accident, in 1914, the rents, royalties, and taxes from Christmas Island had long since completely repaid the British Government for the *Challenger* Expedition. Indeed, Murray had maintained that one was the direct consequence of the other. Disdaining further government help, he had

A rookery of sea birds, the source of an eventual phosphorite deposit.

moved the Challenger Society office to his country mansion, and, like his old friend Alexander Agassiz, he undertook private oceanographic research.

The British Navy charts spanned the world, and if one high coral island was a phosphate mine, why not others? A fateful sample of an odd grey rock was collected about 1897 on the Pacific island of Nauru by a representative of an Australian interisland trading company. The rock found its way to the company headquarters in Sydney, where, three years later, A. F. Ellis (the future Sir Albert) noticed it being used to hold open the door of the analytical laboratory. Geologists said it was petrified wood, but, after three months, Ellis tested it for phosphate. The isolated island of Nauru was politically an extension of the German territory of the Marshall and Caroline Islands, and the mineral rights belonged to a German trading company. However, the Australians had coconut plantations in the Carolines, which the German company wanted to obtain to consolidate its holdings. A trade was made, apparently without any discussion of a certain chemical analysis in Sydney.

The nearest island to Nauru is Ocean, which was a similar high coral platform but iron-bound and ignored. It was annexed to the British empire in 1901 by a landing party from H.M.S. *Pylades.* Exploitation of both islands began soon after, but not until unique loading conveyers were cantilevered out over the open sea.

If three high islands were phosphate mines, why not all? The British and Australians reached Makatea about 1908, but the French were better informed than the Germans had been eight years before. Even so, the British ended up with a substantial interest in the joint company that developed the mines. Accepting Sir John Murray's correlation between basic research and industry, the British obtained a virtual world monopoly of insular phosphate as a consequence of the *Challenger* Expedition. As to present day circumstances, Nauru has the highest per capita wealth of any nation on earth.

The origin of the phosphate rock long remained somewhat of a mystery, but it is now attributed to weathering or leaching of bird droppings. The guano birds of Peru thrive on the teeming fisheries of the cold currents offshore. For protection from predators, the birds prefer to nest on islands. The guano they deposit covers large parts of islands to depths of tens of meters. The total deposit on the west coast of South America is estimated to be 170,000 metric tons per year. If the phosphate rock of elevated atolls was deposited as guano at comparable rates, the great ore bodies could have accumulated in less than a million years.

The accumulation of large deposits of guano depends on satisfying several biological and geophysical conditions. It is necessary, to begin with, that the animals form large colonies that range for food over wide areas but rest and reproduce in a small space. It is required that the excreta remain where it is dropped; this is typical of bird droppings because the urine is semi-solid and contains uric acid, which is relatively insoluble compared with the urea of mammalian urine. Geophysically, two things are required. An arid climate at the breeding and living site is necessary to preserve the guano. On the other hand, the much larger surrounding area that is the source of food must be highly fertile. By far the easiest way to meet all these conditions is to put a large colony of diving birds, such as pelicans, boobies, or cormorants, on an island where the rainfall is low but the sea is cold and fertile because of upwelling. In the Pacific, upwelling and low rainfall occur along the eastern boundary and in the equatorial zone.

All the conditions were favorable for the accumulation and preservation of large quantities of guano on Nauru and Ocean islands, which are in the belt of low rainfall just south of the equator. Makatea remains a puzzle. The rainfall is suitably low, but there is no sign of a bountiful fishery in the surrounding waters except for the phosphate ore itself. For some reason, such a fishery existed in Pleistocene time.

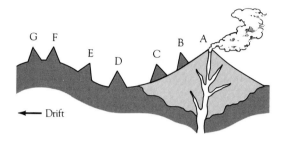

Sequential flexing of a line of volcanoes produced by a hot spot under a drifting plate.

Sequential Loading of Linear Archipelagoes

Atolls may be elevated and subside in complex patterns when a young volcano chances to grow among them. Much simpler interactions may be predicted between the closely spaced islands in a linear progression away from a hot spot. Indeed, these interactions must be commonplace, but evidence that they are occurring is still sparse and conjectural. The essence of the idea can be visualized in terms of a line of closely spaced or overlapping volcanic islands in an age progression from active to oldest. The load of the active volcano A tilts, depresses, and produces normal tensional faulting in island B, which had done about the same things to island C. Farther away, island D may be in the bottom of the moat generated by lithospheric flexing in response to the load of A. Island E was in the moat of B but is now being elevated by the edge of the flexed arch due to the load of A. Islands F and G may be on the crest of the flexed arch, and so on.

The most intensely studied line of islands is the Hawaiian archipelago. Various phenomena there can be explained by sequential loading. With regard to loading at present, all of southern Hawaii is active so the attached northern part of the island should be subsiding both because of overlap and because it is in the flexed moat. In contrast, Maui is on the inner flank of the present arch; Oahu is near the crest; and Kauai is near the outer limit of the arch. Net movement of some islands relative to sea level is shown by tide gauges, which indicate that Hilo, Hawaii, is subsiding at 4.8 mm/year, Kahului, Maui, is subsiding at 1.7 mm/year, and Honolulu, Oahu, is stable. These observations do not appear to support the hypothesis of sequential flexing. However, they give net movement, and in order to identify the component caused by loading, it is necessary to remove the effects of thermal subsidence of the lithosphere.

The Society Islands are a useful place to attempt to separate the two factors. The load of young and active volcanoes in the southern Societies has elevated Makatea more than 100 m in only 1 Ma. About the same distance as Makatea from the center of loading are the barrier-reef islands of the northern Societies. These beautiful Isles sous le Vent include Bora Bora, Raiatea, and Huahine, and none of them have elevated reefs. Presumably, tide gauges would indicate stable or subsiding islands like Maui, which is also at the expected location of an upwardly flexed arch. As the chapter on islands with coral showed, the rate of subsidence can be estimated from the thickness of the reefs and the age of the islands. Raiatea has subsided 320 m to 340 m in 2.4 Ma to 2.6 Ma, Bora Bora 200 m to 360 m in 3.1 Ma to 3.4 Ma. The barrier-reef islands, on average, have

subsided as though they were on lithosphere that has been thermally rejuvenated to an age of only 1 Ma. At such an age, a subsidence of 100 m/Ma
is quite normal. Consequently, even if they were overriding an arch, the
barrier-reef islands would not be elevated relative to sea level. It is interesting to consider that the subsidence of Moorea, 280 m to 340 m, is as
great as that of the Isles sous le Vent even though it is 1 Ma to 1.5 Ma
younger.

The reef thicknesses are approximate but they are consistent with the
hypothesis that the Isles sous le Vent are overriding an arch and thereby
counteracting thermal subsidence, which has greatly affected Moorea.
With this thought in mind, the data from tide gauges in the Hawaiian
Islands can be reexamined. Presumably, thermal subsidence there is more
or less balanced by isostatic uplift in response to erosion. The differential
net movement relative to sea level at Maui and Oahu may reflect either of
these phenomena or overriding of an flexed arch.

Although modern measurements of sea level are ambiguous, geological observations of many kinds are more suggestive. For example, insular
shelves are cut into the slopes of young volcanic islands once they become
inactive. The load of an overlapping active volcano would depress such a
shelf, which would be preserved while a new, higher shelf was cut as the
island drifted away from the hot spot. Exactly such deep terraces have
been observed in the Hawaiian Islands. There is a broad terrace 360 m
deep off Kohala volcano, which was inevitably loaded by the overlap of
the younger volcanoes of the island of Hawaii. Fossil coral reefs have been
dredged from the edges of shelves at about 500 m deep off Oahu and
Lanai. In fact, Hawaiian marine geologists have found a wealth of deep
terraces throughout the islands.

The initial depth of the terraces apparently depends on local loading,
which is quite variable, so the terraces cannot be used as indicators of
overriding an arch. However, volcanism may give such indicators. The
crests of arches around island groups are faulted in some places; this is
usually interpreted as "keystone" faulting caused by the tension that accompanies bending. The crests are also the sites of volcanism. It was for
this reason that I proposed, in 1964, that the youngest volcano in a linear
age progression tended to be located on the crest of the arch of the next
youngest. In those days before drifting plates and hotspots, I visualized a
straight propagating crack intersecting a U-shaped arch curving around
the young end of the line. However, the only one of many active volcanoes that is in about the right position is Macdonald Seamount, and that
may be a coincidence.

It now appears that an arch forms a circle around the latest load,
instead of just a U, so the basic idea of volcanism on a migrating arch that

Top to bottom: Rejuvenated volcanism due to sequential flexing of a series of volcanoes produced by a hot spot.

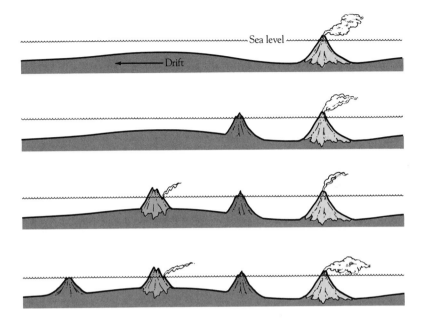

intersects a line of volcanoes can be applied, to use modern terms, downdrift as well as updrift from a hot spot. Visualize a circular arch that stays centered on a hot spot because of the successive construction of volcanic loads. With a constant rate of drift, a rejuvenated phase of volcanism should occur along the line at a constant distance from a volcanic island being built at the same time. Moreover, the lines of tension on the arch ideally should be concentric around the center of loading and parallel to the crest of the arch. Thus, even if there is a bend in the line of volcanoes, the orientation of the rejuvenated volcanism ideally would reflect such lines of tension.

Both the shield-building stages and rejuvenated stages of Hawaiian volcanism have been dated, and the line of volcanic centers bends. Thus, a test of the hypothesis is straightforward. It should be remembered that only the crestal region of an arch is visualized as the locus of rejuvenation. If there is no young volcano at the right distance, about 280 km to 320 km, this mechanism will not produce rejuvenated volcanism on an old island. If this is kept in mind, the hypothesis is strongly supported by the history of Hawaiian volcanism.

Tholeiitic shield building of Kauai took place over a period of more than 2 Ma ending 3.5 Ma ago. The rejuvenation phase took the form of

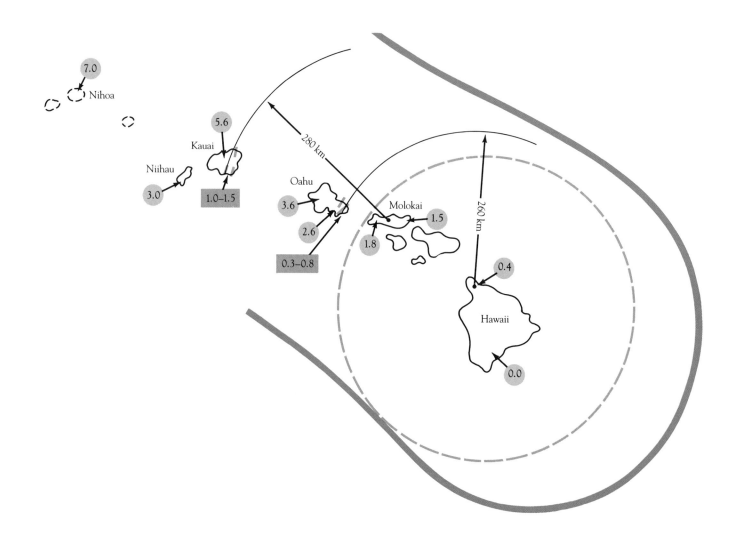

The distance between contemporary centers of primary (orange) and secondary (light blue) volcanism, supports the hypothesis of sequential rejuvenation on a circular arch (light blue circle) centered on a hot spot. The dark blue line represents the crest of the flexed arch that now surrounds the Hawaiian chain.

outpourings of basanites, nephelinites, and related rocks along 40 vents aligned north-northeast. It took place about 1.4 Ma ago; about 1.8 Ma to 1.5 Ma ago, the shield-building volcanoes of Molokai were active. The average distance between the centers of shield building and rejuvenated volcanism is 280 km, and a line between them is very close to perpendicular to the trend of the vents on Kauai.

A similar relation exists between the rejuvenated volcanism of Oahu and the shield building of Kohala. Kohala was built about 0.4 Ma ago; the rejuvenation of Oahu volcanism, with basanites, nephelinites, and the like, was 0.5 Ma to 0.02 Ma ago. The pair are separated by an average distance of 260 km, and the trend of groups of the 30 vents on Oahu is nearly perpendicular to a line to Kohala.

The most southerly occurrence of rejuvenated volcanism is of recent to historic age—it may still be active. It extends along the southern edge of Maui and to the adjacent island of Kahoolawe. The vents are along older rifts that trend southwest and east but not in rifts with a northerly orientation. The general trend of the rejuvenated zone is within a few degrees of perpendicular to a line to the active submarine volcano Loihi, which is 220 km away.

In sum, it appears that a line of volcanoes generated by a plate drifting over a hot spot undergoes sequential flexing.

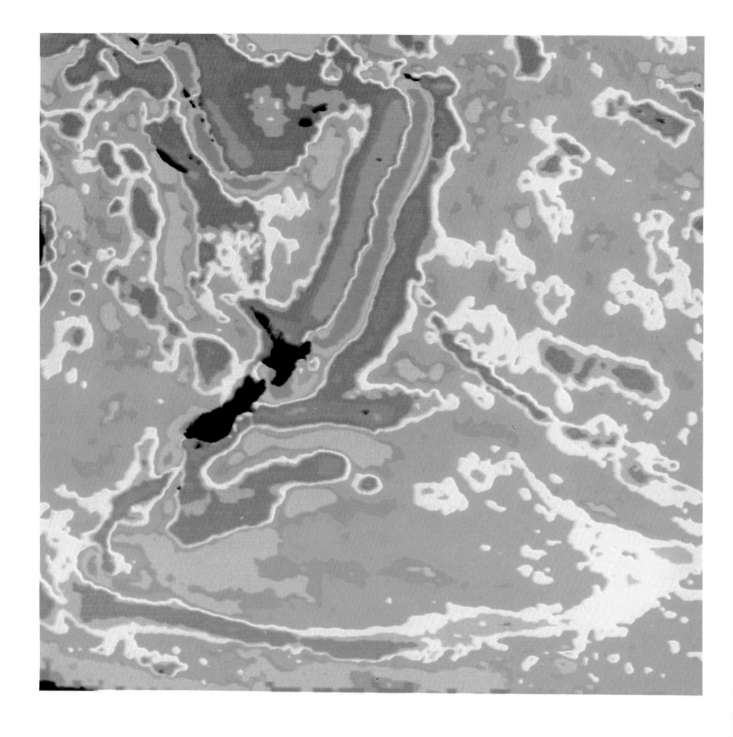

9

Subduction and Accretion

Hot spots constantly generate large volcanoes, yet they do not fill the ocean basins after eons of geological time. The reason is that older volcanoes are disappearing from the ocean basins into subduction zones.

What happens there depends on the age (or thermally rejuvenated age) of the plates, their rate and angle of convergence, their buoyancy, and their homogeneity. For example, if a plate is young, hot, and weak it tends to deform more readily than if it is old, cold, and strong. Likewise, if plates are of equal age but converging at different rates, only rapid convergence produces subduction. Moreover, buoyant continents tend to stick together at the surface when they converge, but the denser ocean floor tends to plunge into the mantle.

The preceding are all examples of the general properties of plates, but the effects of inhomogeneities are more particularly related to the fate of islands. Large clusters of islands are gradational in many geophysical characteristics between continental and oceanic crust. Thus, their behavior in numerous subduction and accretion zones can be revealing regarding the behavior of the few plates that are converging in a small world.

One thing that is clear about modern convergence zones is that the plate boundaries are not closely linked to continental boundaries. At present, there are seismically active convergences between continent and continent, between oceanic plates, and between continent and oceanic crust.

In this computer-processed image of the Pacific near New Zealand, areas of lowest gravity appear dark blue, and those of highest gravity appear pink; land above water appears black. The blue Tonga-Kermadec Trench stretches northeast from New Zealand; near its curved northern end are the Samoan islands. The Louisville Ridge, a thin red strip, points northwest to the center of the trench.

CONTINENTAL CRUST

The collision of drifting continents produces high mountains, great igneous intrusions, deep erosion, thick sedimentary deposits, and other phenomena that are of great interest in geology, but far from our focus on
oceanic islands. However, continental collisions have the important characteristic that the results are buoyant, so they stay at the surface and are
gradually exposed to different depths of erosion. In contrast, the phenomena on one side of continent-ocean and ocean-ocean collisions tend to
disappear by subduction, and on the other side they tend to be concealed
by the sea. Thus, it seems reasonable to outline what is seen in ancient
continental rocks before turning to what is sensed in other ways in modern
oceanic subduction zones.

Continents appear to split relatively easily compared with old ocean
basins, so that Atlantic-type ocean basins form and the crust behind a
drifting continent therefore is oceanic. The leading edges of most drifting
continents also have oceanic crust in front of them, so, before converging
continents collide, the oceanic crust between them must be eliminated by
subduction. The leading edges of drifting continents have shapes that
reflect how they were formed, whether by earlier splitting, accretion of
smaller fragments, or subduction of oceanic crust. In any event, it is highly
unlikely that colliding continents will have perfectly congruent margins.
Instead, two randomly irregular margins are pushed together, putting the
projecting peninsulas under intense stress while the intervening bights or
bays are not yet in contact. This has happened where Africa has been
colliding squarely with southern Europe. The result has been that the
peninsulas tend to be sheared off and moved laterally to fill the bights.

Different phenomena might be expected to accompany a diagonal
collision. Such a collision apparently has greatly affected or wholly dominated the Mesozoic-Cenozoic geological history of western North America, much of which consists of crustal blocks, many of them quite unrelated, that are bounded by faults. Their source is uncertain, although
perhaps it was a lost continent, "Pacifica." Some real botanists and pseu-
doanthropologists appeal to the former existence of now lost continents to
explain data in their subjects. However, the common conjunction of the
words "lost" and "continents" in many scientific disciplines should not be
taken as independent support for the concept. Quite the contrary. The
botanical and anthropological concept concerns temporary continents
that sank beneath the ocean in geologically recent time. The geological
concept generally refers to continental crust that is fragmented and drifts
about while still above sea level. Only rarely and after geologically long

The fragments of Nur and Ben-Avraham's Pacifica drifted across the Pacific for 115 million years to collide with Asia and the Americas. Part A, 225 Ma ago; B, 180 Ma; C, 135 Ma; D, 65 Ma. Fine lines indicate the present-day continental outlines.

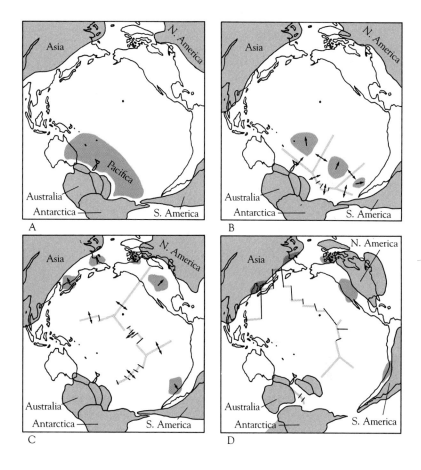

periods do deep erosion and thermal subsidence manage to combine to pull a whole continental fragment under water.

As visualized by Amos Nur and Zvi Ben-Avraham, Pacifica was once draped around the northern margins of Australia and Antarctica when they themselves were joined. By 180 Ma ago, Pacifica was split into at least three large fragments, which by 65 Ma ago were destined to collide respectively with Asia and North and South America. Wherever they came from, the fragment or fragments that reached North America were broken into small pieces and rotated as they drifted northward in a diagonal collision zone.

A third type of collision apparently has occurred where India has pushed deep into southern Asia and produced high mountains on three sides of the sub-continent. The crust under the central Asian plateaus

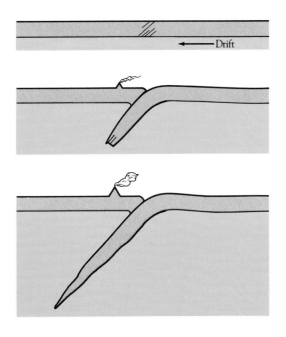

Top to bottom: Stresses in an old drifting plate lead to breaking; the broken edges sink to form a trench; one of the plates descends into the trench and begins to plunge into the lithosphere.

north of the Himalayas is extraordinarily thick. A simple explanation for the thick crust and high plateaus is that the leading margin of the Indian continent was pushed down and under Asia so the two now overlap even though both are buoyant.

The sutures that form between continents in collision tend to be as strong or stronger than the interiors of the continents. This is evident in the geology of the continental margins of the North Atlantic. A proto-Atlantic existed at one time but closed up before the present period of drifting began, roughly 200 Ma ago. The older suture did not reopen everywhere, although it was not distant from the new crack. As a consequence, fragments of pre-suture Europe and North America were left stranded amid alien terrain to puzzle generations of geologists.

In sum, the lithosphere breaks up in very complicated but generally predictable ways where continents collide. Fragments may move laterally, rotate, and even overlap, but, because of their buoyancy, they apparently do not sink into the mantle.

OCEANIC CRUST

A simple model for the start of subduction starts with old, cold, drifting oceanic lithosphere about 100 km thick. For some reason, a line of convergence and compression forms within the plate and becomes connected to other such lines by transform faults. As strain builds up the plate breaks; the broken edges sink, because the plate is denser than the mantle, forming an oceanic trench. As the plate continues to move forward, its leading edge begins to descend into the trench. The dip of the sinking slab is commonly 45 degrees or so; thus the rate of subsidence is about the same as the rate of drift—50 mm/yr to 100 mm/yr. The sinking part of the slab, bending downward, produces a balancing outer arch parallel to the trench.

The plunging lithosphere is not only denser than the surrounding mantle, it is also more rigid and colder. What happens resembles the slow immersion of a slab of cold solid wax into a vat of boiling wax. The edges and sides of the slab begin to melt as soon as they are immersed. After a time the leading edge is completely melted and the immersed solid slab has an approximately triangular longitudinal section. As the slab is heated, its rigidity decreases. Nonetheless, the slab retains its integrity at least to a depth of 700 km. This is known because of the obvious earthquakes it generates during subduction. Seismological tomography makes it possible to detect anomalous regions in the mantle by comparing the behavior of earthquake waves along different paths. By this method, T. H. Jordan and his colleagues have traced subducting lithosphere below 1000 km. Thus, the length of subducted slab may be as great as 1400 km. At drift rates of 50 mm/yr to 100 mm/yr, some lithosphere may continue to exist in the mantle for times in the range of 30 million years.

One of the most interesting characteristics of a subducting slab is that it is capable of pulling oceanic lithosphere toward a subduction zone. Visualize a sheet of paper that is hanging over the edge of a table. If enough paper is hanging, gravitational attraction overcomes friction and the paper slides off the table. Exactly the same thing happens once a part of the lithosphere breaks free and begins to sink. At least during Tertiary time, this *trench pull* has dominated the forces that drive drifting plates. Those, like the Pacific and Nazca one, that are being pulled into trenches drift rapidly. In contrast, the American and other plates that are not being subducted drift very slowly.

Effects of Age

Most of the phenomena associated with the subduction of normal oceanic crust are influenced by the age of the lithosphere. For example, the outer arch may be almost 1000 meters high (although still submerged) where crust 100 Ma old bends into a trench, but undetectable if the crust is only 10 Ma old. Presumably, the difference in depth of two converging plates has an influence on which one will be subducted. Another factor that varies with age is the thickness of the cool lithosphere. Consequently, the time required to reheat and destroy a plunging slab also varies with the age of the lithosphere.

So does the rate of trench pull, because a young, thin, short-lived slab is more buoyant than an older one. This age dependence has a curious consequence when subduction is diagonal to the direction of sea-floor spreading. Such diagonal subduction has been typical of the intersection of the East Pacific Rise with the North American plate. As the east flank of the rise is being subducted, some sections break free from the rest of the east flank by newly created ridge-trench transform faults. Then the new, more or less triangular plates continue to move toward the subduction zone, with the older lithosphere being pulled faster than the younger—the hot, buoyant ridge crests acting drogues to slow and guide the drift of the plates. The end result is that the new plates rotate into the trench until the ridge crests are parallel to and near the subduction zone. Then the trench pull is equal everywhere and rotation ceases. The effects of pivoting subduction are readily seen in the arcuate fracture zones of the East Pacific Rise off Middle America (next page). The phenomenon of pivoting subduction shows the large tectonic effects that may result from small variations in the buoyancy of plates.

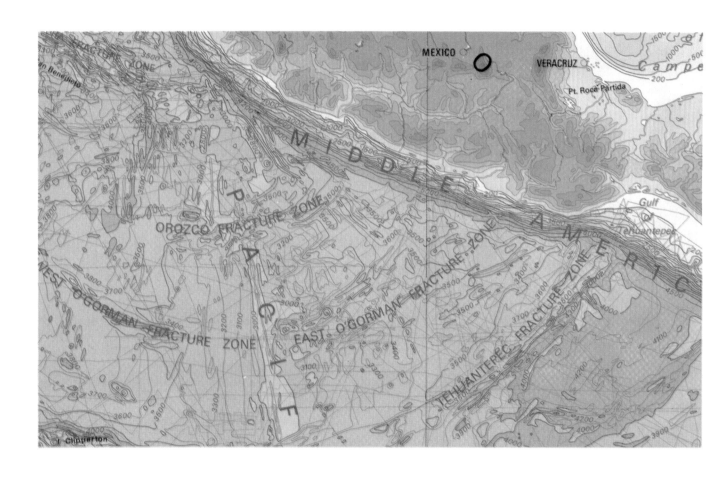

Curved fracture zones off the Pacific coast of Mexico indicate that plate fragments are rotating to bring segments of the East Pacific Rise into alignment with the subduction trench.

Obduction

In the 1960s, geologists were excited by a proposal to drill through the earth's crust and sample the mantle. The bottom of the crust was thought to be the Mohorovicic discontinuity, or "Moho"—a seismically detected surface at an average depth of 32 km under continents and 10 km under oceans. The proposal was to drill at sea. Preliminary sea trials successfully drilled into the top of the oceanic crust, but an enormous increase in Federal funding was needed for further drilling. The program was cancelled, and it appeared that the mantle would remain beyond human grasp.

Within a decade, the new concepts of sea-floor spreading and plate tectonics revolutionized geological ideas. It became possible to construct detailed stratigraphic models of the oceanic crust and upper mantle. It soon was realized that the exact sequence of distinctive igneous rocks

The ophiolite suite of rocks formed at spreading centers is characteristic of oceanic crust.

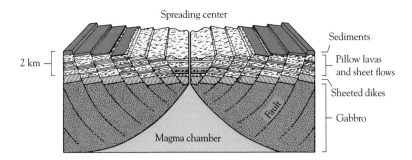

predicted by some models was exposed on land. It was the unusual igneous sequence that the indefatigable field geologists had recognized and given the name of *ophiolite suite.* Ideally, a simple oceanic crust consists of the following sequence from top to bottom: pillow lavas formed by submarine lava flows; sheeted dikes that fill in near-surface cracks in spreading centers; and coarsely crystalline rocks called *gabbro* that solidify in a magma chamber below the spreading center. Above this sequence might be lithified pelagic red clays or calcareous oozes; below it would be the dark green rocks of the uppermost mantle. Ophiolite suites, more or less complete rock sequences from pelagic sediments to the mantle, are found in the Samail complex in Oman and the Blow-Me-Down complex in Newfoundland.

Thus, it is not necessary to drill at sea and at great expense to sample the mantle. It can be done with a geology pick, because sometimes the oceanic crust is *obducted,* it overrides continental crust, instead of subducted. However, the rarity of ophiolite outcrops shows that obduction may require very unusual circumstances. For example, R. G. Coleman has proposed, regarding the Samail complex, that a spreading center died and the still hot oceanic crust was compressed and fractured along oblique thrust faults in Middle Cretaceous time. This complex of superimposed blocks of oceanic crust was obducted onto carbonate rocks of the continental shelf in Eocene time. It is not certain which characteristics of the crust caused it to be obducted instead of subducted. However, the water depth was relatively shallow because the crust was young, and that would have favored obduction.

Underplating

Thin, relatively buoyant oceanic crust may be obducted onto a continent. Normal dense oceanic crust is subducted into the mantle. If the oceanic crust is almost neutrally buoyant relative to the mantle, it may be capable of undercoating another oceanic plate, just as one continental plate apparently has been pushed under another in central Asia. Some observations of subducting seamounts suggest that underplating of oceanic crust is indeed an important process.

EDGES OF UNSUBDUCTED PLATES

In many places, oceanic plates have been subducted for thousands of kilometers. Almost inevitably, such prolonged action by one plate in a subduction zone effects profound changes on the edge of the other, more passive plate in the zone—whether it be oceanic or continental. These changes include volcanism, tensional stretching, and accumulation of offscraped and obducted sediment.

The volcanism is of a type characteristic of subduction zones. Andesitic and basaltic lavas build regularly spaced single lines of volcanoes parallel to oceanic trenches. Such volcanoes form the high, symmetrical, beautiful peaks of the Andes (whence the name andesite) and the Cascade Mountains on land, and the more numerous volcanoes of island arcs. Andesite is varicolored and characterized by more acidic minerals than basalt. A close association exists between the depth of the plunging lithosphere and the composition of igneous rocks extruded or intruded in the plate above. It appears that melts derived from the lithosphere are mixed with the overlying rocks to produce the andesitic volcanoes from a source depth of about 125 kilometers. It should be noted that, although the lines of andesitic volcanoes are exceptionally long and regular, there are gaps. The gaps, like many other unusual local variations in subduction zones, seem to be due to the subduction of oceanic volcanoes.

Charles Darwin thought he discovered an association between uplifted coral islands and active volcanoes. It was in island arcs rather than the interiors of oceanic plates where there is such an association, but that one was unknown to him. The uplift of coral islands in subduction zones is not due to the load of the volcanoes as it is, for example, in Polynesia. This is clear because, viewed in detail along an island arc, uplifted islands are at least as likely to occur near volcanic gaps as not. Vertical movement might occur for many reasons in a tectonically active subduction zone, and it is difficult to isolate the cause. Whatever it was, it did produce large, flat-topped limestone islands in an enormous region otherwise populated with watery atolls and steep-sided volcanoes. Among these were the Marianas Islands of Guam, Saipan, and Tinian, which, after American recapture in World War II, became enormous aircraft carriers. These were not uplifted on the outer arch but between the trench and volcanic arcs.

Back-arc basins are another feature generated at the margin of a relatively passive plate by extensive subduction of an adjacent plate. The basins are parallel to trenches and island arcs or their continental equivalents, lines of andesitic volcanoes. The Japan Sea is an example. At sea, the basins are relatively shallow—about the depth of the crest of a spreading midocean ridge. They are seismically active, although rarely along discrete lines like a midocean ridge crest. There are numerous indicators that the floors of the basins are young. Sediment is generally thin and exposed, rocks include fresh, glassy basalt. Magnetic anomalies like the oceanic type appear to exist, but they are usually jumbled or disorganized and cannot be used to prove symmetrical spreading or the age of the crust in the basin. Nonetheless, spreading centers have been identified by the relative degree of weathering of igneous rocks and the thickness of overlying sediment. It appears, therefore, that back-arc basins are produced by a kind of secondary sea-floor spreading. That being so, the lithosphere of the island arc between a back arc basin and a trench is, in effect, a long, narrow, but separate plate moving between two large ones. As the basin widens, the island arc moves toward the trench. Considering that some back-arc basins are hundreds of kilometers wide, it appears that the trench may retreat before the advancing island arc. Such a retreat might be expected, considering that the lithosphere is unstably denser than the mantle. Even without drifting, a plate would be subducted as a subduction zone migrated toward it.

Offscraping and related phenomena take place on a large scale in subduction zones because of the enormous volume of sediment involved. For example, if pelagic sediment averages 200 m thick, subduction of only 5 km of oceanic lithosphere will push sediment 1 km thick and 1 km wide into the length of the oceanic trench. A mere 1000 kilometers of subduction piles up 200 km^3 of sediment per kilometer of trench. One of the many things that can happen is readily seen in multi-channel seismic profiles of the Aleutian Trench and island arc. Sand, mud, and volcanic ash from Alaska and the Aleutian Islands accumulate in the trench. The subducting Pacific plate pushes this "hemipelagic" sediment plus the pelagic sediment from the open sea against the edge of the North American plate. The edge acts as a stationary, horizontal chisel, which splits the sediment pushed against it. The sediment detached above the chisel is obducted onto the margin of the North American plate. The sediment below the chisel remains attached to the plate as it begins to plunge. However, the seismic profiles suggest that at least part of the lower sediment is underplated on the continental margin instead of being subducted with the oceanic plate.

A common feature of older island arcs is a bench some tens of kilometers wide between the trench and the uplifted older islands or active volcanoes. The bench has been studied by seismic profiling and deep-sea drilling; the more information that is collected, the more complicated the

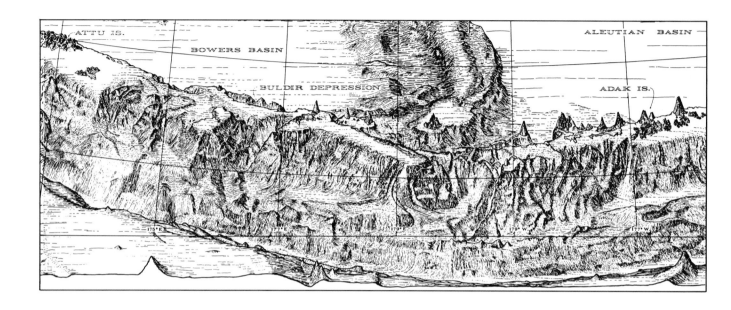

ATTU IS.   BOWERS BASIN   ALEUTIAN BASIN

BULDIR DEPRESSION   ADAK IS.

View towards the north of the Aleutian Trench. The bench on the far wall consists of oceanic sediment and volcanoes scraped off the Pacific plate as it plunges into the trench.

structure appears. The bench clearly consists of sediment, sedimentary rock, and volcanic rock that has been obducted onto the inner, or relatively passive, side of the trench.

The concept of obduction was difficult to accept when it was first proposed because it requires that relatively weak materials be pushed up against the force of gravity and yet retain some coherence in blocks. Faulting, rotation, and intrusions apparently occur, but the end result is obduction. This process, which can be seen in action in contemporary subduction zones, seems to explain much of what is known about the geological formations called "melanges." Such a one is the Franciscan formation in California, which consists of small slices and bits of rock typical of oceanic crust and deep-water pelagic sedimentary rocks. In Mesozoic time, these rocks were emplaced in a subduction zone that no longer exists.

ISLANDS IN SUBDUCTION ZONES

The fate of islands that drift into a subduction zone depends on their size, thickness, strength, density, and rate and angle of approach. These factors determine whether oceanic islands are fragmented or preserved whole, and whether they are subducted, underplated, obducted, or plastered onto the

These tortured rocks of the Franciscan formation near San Francisco were scraped off a subducting oceanic plate and plastered onto North America.

edge of another plate. At this point, it seems appropriate to broaden our subject from islands made either of coral or volcanic rock. All kinds of islands arise from deep water, and they exemplify a wide range of the factors that determine what happens to islands in subduction zones. We shall deal now with four types of islands: isolated volcanoes, volcanic plateaus, tiny continents, and the large fraction whose crustal structure is in doubt.

As to isolated volcanoes, little can be added to what has already been said. Volcanic islands tend to be less dense than deep-sea basalts because of the inclusion of vesicles. Thus they are relatively buoyant as well as isostatically compensated either locally or regionally. Concerning their structural integrity, oceanic volcanoes are vulnerable to slumping, faulting, and radial cracking by intruding magma. They have no lateral supports. It appears that they should break up rather easily in subduction zones.

Oceanic plateaus differ from isolated volcanoes chiefly in size. The two great modern examples are Iceland and the Galapagos Islands. Each is characterized by a dozen or more closely spaced active volcanoes whose bases overlap to form the plateau. At Iceland, the sea floor is shallow enough and the volume of lava flows is great enough to build a plateau above sea level surmounted by volcanic mountains. In the Galapagos, the plateau is a shallow shelf above which rise volcanic islands. The islands are most abundant on the western end of the plateau above the presumed hot spot. Farther east, on the hot spot trace, the shelf deepens because of thermal subsidence. The older islands were truncated by erosion while the shelf sank. Consequently, the deep shelf is dotted with broad banks only a few hundred meters high. In due course they will become closed spaced, low guyots. The circumstances in Iceland are different in that the whole top of the vast volcanic plateau is exposed to leveling by erosion and deposition. Rather than many broad, low banks, one enormous bank is being produced. Deep-sea drilling has confirmed that the deep, relatively flat-topped submarine ridge east of Iceland is an enormous guyot derived from the Iceland hot spot. Even on the scale of Iceland, the sequence is confirmed: islands become first banks and then guyots or atolls.

Numerous drowned ancient plateaus capped by guyots have been identified. Satellite gravity observations indicate that most are compensated relatively locally, indicating that they grew as loads on the young crust of midocean ridge crests. The fact that the guyots of the plateaus were truncated in water 2.5 km to 3.5 km deep—typical of ridge crests—is shown by their relief above the surrounding sea floor. These and many other kinds of observations indicate that the voluminous volcanism in relatively shallow water that produces oceanic plateaus inevitably pro-

The Galapagos Islands rise from an oceanic plateau.

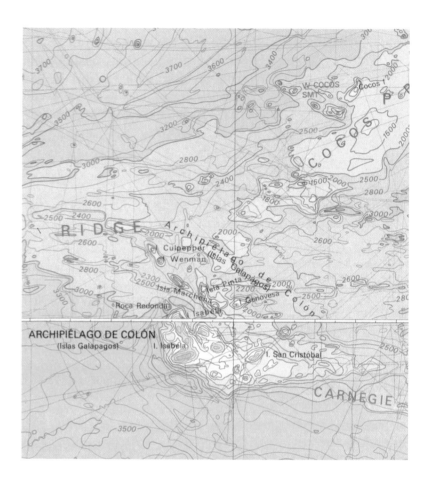

duces islands as well. Indeed, it would be difficult for submarine volcanism to be so delicately balanced as to build a vast plateau without at least some peaks that become islands. There is, however, a strong suggestion of a near approach when a broad plateau is just below sea level. This can be seen in the extraordinary abundance of very low, broad guyots that rise only a few hundred meters above the main plateaus of the Mid-Pacific Mountains and the Tuamotu Ridge. Somehow, despite the number of volcanoes, volcanism persisted in the style of the Galapagos instead of going to the style of Iceland.

The crustal structure of oceanic volcanic plateaus is obscure at present. Many attempts have been made to measure crustal sections, but explosion seismology is particularly difficult over mountains and plateaus. Moreover, in some places where sections have been successfully measured, the results are difficult to compare with standard continental or oceanic crust. Thus there appear to be only a few places where the plateaus are unquestionably volcanic and where something is known of the structure. Iceland has about 10 kilometers of volcanic rock over "anomalous" mantle, which may also be in part the result of local igneous intrusions. Ninety-East Ridge, in the Indian Ocean, has similar structure. The Nazca Ridge has a rather normal sequence of layers for oceanic crust, but the total thickness is about triple normal.

At the opposite extreme from plateaus that are certainly volcanic are those that are certainly continental. Such "non-magnetic seamounts" are not uncommon around the margins of the deep North Atlantic. Examples are the Rockall Plateau and Porcupine Bank. Some have outcrops of granite or other continental rocks, and others have crustal sections resembling continents. They would give magnetic signatures if they were volcanic, and they do not. There is little doubt that these small plateaus rising from the deep sea are merely detached slivers or fragments of Europe or North America. They were isolated mostly during the interval when the location of the Atlantic spreading center was still under adjustment.

Although the North Atlantic fragments are presently relatively close to the mother continents, at other times fragments have drifted about and become completely separated. The most conspicuous example at present is the Seychelles Bank in the Indian Ocean. These famous islands consist of a core of coarsely crystallized granite ringed with coral reefs. The granite is about 700 Ma old and the plateau has been eroded deeply enough to expose it; the crustal section is completely normal for a continent.

Despite the prodigious efforts of marine geophysicists during the past 35 years, the origin of a large number of submerged oceanic plateaus or rises capped with guyots remains in dispute. Ship-borne seismologists have used explosions to measure the crustal structure of most of the major plateaus, but their facts can be interpreted in different ways. For example, in the Pacific are the Ontong Java, Manihiki, and Shatsky rises. The crustal layers under these rises have about the same seismic velocities as the layers in normal oceanic crust. However, the thickness of each layer under the Ontong Java Rise is five times normal, and for the other two rises it is about three times normal. Thus, D. P. Hussong and colleagues reason that the rises or plateaus are indeed oceanic crust produced in the normal way but in greater quantity than usual. On the other hand, Nur and Ben-Avraham have observed that if the layers of normal ocean crust are similarly multiplied by a factor of about five, the resulting structure is not very

different from normal continental crust. Thus, there remain many pla-
teau-like fragments of crust in ocean basins whose origin is uncertain.

It is evident that large and small, deep and shallow, and thick and
thin fragments of crust drift into subduction zones, but not even the proportions of continental and oceanic fragments can be determined at present. The subduction, obduction, and collision of such fragments in the
past has produced the present topography and structure of oceanic
trenches. Consequently, the interpretation of observations of islands,
atolls, and guyots that are now in subduction zones is clouded by much
more uncertainty than interpretations of mid-plate phenomena.

Uplift on a Flexed Arch

At least ten former atolls are now positioned on an outer arch that is
produced by upward flexing of the lithosphere before it plunges into an
oceanic trench. Others may now be uplifted on an arch, but they are in
tectonically complex regions and the cause is uncertain. However, no
atoll in the world that is presently on an arch is not uplifted. Thus, it is
a reasonable conclusion that not only every atoll but every high island,
guyot, or seamount that drifts over an arch is also uplifted.

The amount of uplift ranges from about 20 m to 355 m—at Christmas Island in the Indian Ocean. However, the uplift of each island presumably varies as it overrides the arch, reaching a maximum at the crest
before the island slowly disappears into the trench. No single atoll has
been followed through the history by drilling, although it certainly would
be possible, for example, at Ulithi, the World War II carrier task force
anchorage just east of the Palau Trench. In the absence of historical data,
the geologist falls back on the assumption that the present uplift of different atolls may be comparable to the sequential uplift of one. It appears
that this assumption is correct, because the amount of uplift of each atoll
varies approximately with its distance from a trench axis in conformance
with the shape of the arch.

Given that individual atolls override arches, it is possible to reconstruct an idealized history. Arches produced by flexing are a few hundred
kilometers wide, and the lithosphere drifts over an arch, like a standing
wave, at 50 mm/yr to 100 mm/yr. Taking, for example, an arch 500 km
wide and a drift rate of 50 mm/year, an atoll would be uplifted for 10 Ma
while overriding. Even for the fastest drift and a narrow arch, the period of
uplift would be a few million years. Thus, atolls lifted by this cause tend to
be exposed much longer than midplate atolls uplifted by the load of young
volcanoes. Such a long period of exposure should result in conspicuous
erosion, and this can be seen at Christmas Island—the one that made Sir

John Murray rich. The island was once an atoll, but it has been deeply eroded by runoff of rain and solution by groundwater. As a consequence part of the volcanic core of the atoll has been exposed. In addition, waves have cut benches and nips at different levels in the steep sides of the island.

Similar nips occur on other uplifted atolls but the profound solution seen at Christmas Island has not been noted elsewhere. Perhaps this is because rainfall varies markedly in the tropics and there are not many uplifted atolls to observe. Moreover, there is not a single emerged guyot on an arch in temperate waters anywhere at present. Indeed, they may always be rare because of wave erosion. Coral growth protects the flanks of emerging atolls in the tropics, but there is no protection in higher latitudes. Thus, even if a guyot were elevated to sea level, waves might keep planing it away to a shallow bank.

In addition to what is known of modern islands, fragments of the history of two ancient islands on arches are known. One of these is Capricorn guyot just east of the Tonga Trench in the southwest Pacific. My colleagues and I discovered and surveyed it on the *Capricorn* Expedition in 1952. It is right at the edge of the trench and thus has already drifted over the arch. The height of the arch directly east, away from the trench, is obscure, but farther north it appears to be about 700 meters. One reason it is obscure is that a cluster of volcanoes is on the east flank of the arch. All are submarine but Nine Island, which is a former atoll that is now uplifted about 700 meters as it begins to override the arch.

Capricorn guyot has a largely truncated summit, 18 kilometers in diameter, that slopes one degree westward toward the trench and is 800 m to 1000 m deep. At the eastern edge of the main platform is a flat-topped, westward-sloping knoll at a depth of about 400 meters. The guyot was successfully dredged by New Zealand oceanographers in 1958. From 880 meters, they obtained fossils of solitary coral, brachiopods, mollusca, and foraminifera but no evidence of reef coral or other reef biota. Probably, this guyot was never an atoll. Judging by its relief of 4400 meters, it was truncated in cool water when on normal crust with an age of 25 Ma or, more probably, on a midplate swell. The guyot is in tropical waters now, but for some reason it never became an atoll when it apparently was subjected to erosion while overriding the arch. Perhaps it was exposed to waves only briefly during periods of very low sea level during ice ages.

The other guyot of interest is Giacomini, which is in the Gulf of Alaska just east of the Aleutian Trench. The arch east of this trench has been buried locally by deep turbidites (sediment deposited by turbidity currents) derived from southeastern Alaska. However, a broad gravity anomaly shows that the arch exists here, and it can be seen as a topographic feature elsewhere next to the trench. Giacomini guyot is on the arch as indicated by the gravity anomaly. The guyot is 21 Ma old and rests on crust 43 Ma old. It is the oldest dated member of a line of guyots that were generated by two or more hot spots. The shelf break of Giacomini is at only 730 meters, which means that its net subsidence has been only that much instead of the much greater amount expected considering the age of crust and volcano. The younger volcanoes in the line to the southeast have had a greater net subsidence by several hundred meters. It appears, therefore, that Giacomini has been uplifted by at least 200 meters by overriding the arch. Thus, the evidence of guyots as well as atolls confirms that uplift, although not necessarily emergence, universally accompanies the overriding of flexed arches.

The Samoan Islands and Subduction

The Samoan islands in the southwestern Pacific are in a line with the correct orientation to have been generated by the present motion of the Pacific plate over a fixed hot spot. However, the age progression appears to be backward, with a very active volcano on Savaii at the northwestern end of the archipelago; Rose atoll lies at the southeastern end. This progression aroused no concern until it was demonstrated that the whole Pacific plate is rigid and drifting westerly or northwesterly. Then the Samoan age progression became a major anomaly and geologists made new, detailed studies to explain it.

At present everything seems satisfactory in that the data not only explain the age progression in detail, but also illuminate other aspects of plate tectonics. There are no age measurements from Rose atoll, but the fact that it is an atoll suggests that it is relatively old. However, old and young islands are mixed together in most archipelagoes. Thus, it is reasonable enough to interpret Rose as an old island that just happened to drift to where it is. Once Rose atoll is eliminated, all is well. Not far to the northwest is an active submarine volcano and farther northwest along the line are islands with tholeitic shield-building lavas in an age sequence from about 1 Ma to 3 Ma.

What is misleading is the secondary, rejuvenated phase of volcanism that emits alkalic basalts and related rocks. In Hawaii, this phase took place in an age progression a few million years younger than the voluminous shield-building stage and may be a side effect of sequential loading. In Samoa, however, simple sequential loading cannot explain what is happening. Instead, any effects of loading are obscured by voluminous rejuvenated volcanism along a rift zone that extends for 300 kilometers across three islands.

Craters showing recent volcanic activity on the Samoan islands lie along a well-defined rift zone.

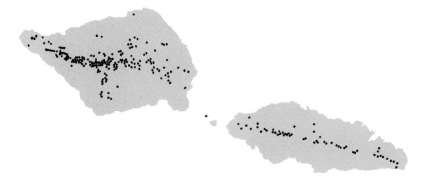

This unusual rift appears to be associated with the subduction of the northern end of the Tonga Trench (see the next page). A basic concept of plate tectonics is that a transform fault leads from the end of a trench either to another trench or to a spreading center. Topographically, the ideal northern Pacific plate drifts serenely past the southern part of the plate, which is going through the convulsions of flexing and subduction. However, the flexing produces an arch several hundred kilometers east of where the Pacific plate actually splits at the end of the trench. Consequently, the adjacent part of the unsubducted plate must also be somewhat elevated. Moreover, the trench does not terminate abruptly at a transform fault but bends gradually to join it. Thus, some sort of flexing might be expected on the unsubducted plate next to the bend. Any such flexing is concealed by the volcanoes of the Samoan archipelago. However, the 300-kilometer-long volcanic rift is roughly parallel to and 100 km to 150 km from the bend. It seems probable that the rift is a manifestation of a flexed arch that bends around with the trench. The small world contains no examples of shield-building volcanism along a trench-trench or ridge-trench transform fault. Indeed, there is only the Samoan example of rejuvenated volcanism superimposed on shield-building generated by a hot spot. However, the unique Samoan data indicate two important facts. Flexing can curve around the end of a trench, and a second phase of volcanism can be triggered by flexing whether caused by subduction or sequential volcanic loading.

A question immediately arises, does flexing next to trenches trigger primary volcanism? The answer appears to be no, because there are no active shield-building volcanoes on the flexed arches. The flexing merely improves the conditions for rejuvenated volcanism along the older channels established by shield building.

The position of the Samoan islands near the curved
northern end of the Tonga Trench suggests that their rejuvenated volcanism is due to flexing of the Pacific plate as
it plunges into the trench.

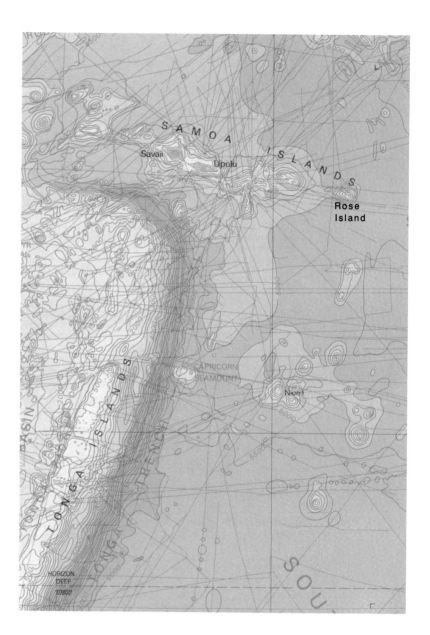

Individual Volcanoes in Trenches

Once the flexed lithosphere has crossed the crest of an arch and is drifting down toward a trench, it enters into a new regime. Recent studies have shown that the outer slope of the trench—toward the arch—is broken by little steplike fault scarps. The down side of most steps is toward the trench. At one place in the Marianas Trench such fault scarps can be seen to offset the lower flank of Fryer guyot. However, surveys with multibeam echo-sounding systems utilizing computerized contouring are necessary to see such detailed topography. Few such surveys have been published, and the steps are not large enough to be seen by conventional sounding techniques. However, it is probable that all volcanoes, ridges, and plateaus are faulted into little steps as they approach the axis of a trench.

Two guyots that lie on trench axes have been mapped by conventional methods. Russian oceanographers have found that Ozbourn guyot has a flat top 15 km to 20 km across at a depth of 2000 m to 2500 m. The volcanic structure is so large that it interrupts the continuity of the Tonga-Kermadec Trench as a topographic feature. Japanese oceanographers have found an island that has been pulled down even farther by subduction. Daiichi-Kashima guyot has a flat top at a depth of 3700 meters, although it rises high above the 7500-meter depths of the axis of the Japan Trench.

Guyots are passive when they are pulled down into trenches, but living atolls may fight back. There are no known examples of the phenomena at present, but quantitative data suggest it is possible. Suppose the lithosphere is drifting toward a trench at about 100 mm/yr. The sides of trenches have slopes between 2° and 7°, so the subsidence into the trench is 5 mm/yr to 10 mm/yr. Charles Darwin showed that healthy coral can grow upward 30 to 100 times faster than that rate of subsidence. Thus, an atoll can easily grow upward to match subduction. However, it cannot maintain its size. The uppermost side slopes of atolls are commonly steeper than 45°, and for the top 2000 m the slopes are 25° to 35°. Thus, the area of an atoll decreases as it subsides. Take Ulithi atoll, which is drifting toward the Palau Trench. The atoll is roughly 10 km wide and 35 km long. Considering only the width, the atoll can grow upward 2.5 km to 5.0 km or so before it comes to a point. Thus, if it maintains its health, the atoll is quite capable of clinging to the sea surface almost until it reaches the axis of the trench. Moreover, Ulithi is not a big atoll—it just happens to be near a trench. Other atolls are 40 km to 50 km in diameter. If one of these went into a trench, it might even continue to be an atoll as its base was sliced off by subduction. Under those circumstances, a structure with reef corals 5 to 7 kilometers thick, no volcanic base, and very low density might rise above the axis of a trench. It seems likely that it would be obducted eventually and appear in the geological record. It also seems that if atolls intersected a trench very often, the continuity of the trench would be destroyed. Fortunately for trenches, few atolls survive the long drift toward the trenches. Instead, they die and become passive guyots.

Volcanic Ridges and Plateaus in Subduction Zones

Large volcanic ridges and plateaus are relatively buoyant elements of the lithosphere, and this characteristic produces various effects when they enter a subduction zone. In general, the effects can be correlated by the simple hypothesis that the relatively buoyant elements do not plunge steeply. Rather, they drift with the oceanic lithosphere almost horizontally just below the lithosphere of the inner, or relatively stationary side of the trench. The effects include the shaping of trenches, the formation and suppression of back-arc basins and active volcanoes, seismicity, and flexed arches.

The most conspicuous and widespread effect of the intersection of trenches with ridges and plateaus is the shaping of the long smooth arcs of trenches. Many ridges intersect the cusps where two arcuate trenches meet. The association is too common to be a coincidence, so it is evident that the ridges and trenches interact. The ridges have not been trapped in pre-existing cusps, because the resulting rotations of the Pacific plate have not occurred. Thus, the ridges have created the cusps. P. R. Vogt and colleagues have proposed that, at ridge intersections, a trench is relatively immobilized. Between intersections the trenches drift slowly toward the plunging plate and create back-arc basins. Gradually, a once straight trench with a continuous back-arc basin may in this way become altered to a chain of arcs with intermittent back-arc basins.

Ridges and plateaus may be obducted because of their low density, but they do not fill trenches as much as they would if all were obducted. Apparently most of these features are subducted—particularly, their roots. What then happens depends on the angle of intersection, especially for a long thin ridge. The phenomenon can be seen in the western Pacific. If such a ridge intersects a trench at right angles, a gap appears in the seismicity of the subduction zone where the ridge apparently extends under the immobile plate. This line of extension is also commonly a boundary for provinces with and without active volcanoes, uplifted coral islands, and back-arc basins. It is possible that all these effects are related to radial fragmentation of the lithosphere. Possibly, the slabs on each side of the intersection are plunging at different rates even though they are still attached to the Pacific plate.

The phenomena associated with the diagonal subduction of a linear ridge are admirably illustrated in the Tonga region. It should be noted that

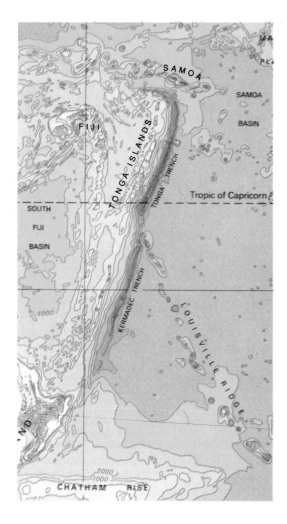

The Louisville Ridge intersects the cusp between the Tonga and Kermadec trenches.

the evidence is circumstantial because the dimensions of any lithospheric anomalies that may have been subducted cannot yet be determined directly. The Pacific basin contains many large volcanic ridges and plateaus and also anomalously thick and shallow regions of unknown origin. It cannot be established that similar features have not been subducted in the Tonga region. However, it is possible to correlate all the observed characteristics of the region with the subduction of a single volcanic ridge.

The Louisville "ridge" is a line of very closely spaced or overlapping volcanoes, including Ozbourn and many other guyots. The ridge was generated by a hot spot on the upper western flank of the East Pacific Rise. The hot spot clearly is a major, long-lived feature at least 40 Ma old. Thus, it is reasonable to assume that the western end of the Louisville ridge has been subducted where it disappears in the Tonga Trench. Moreover, from unsubducted hotspot tracks, the motion of the Pacific plate is known, so the subducted part of the ridge is known to have a trend that is more northwesterly than the present direction of plate drift in the region. Taking the north-northeasterly trend of the trench into account, the ridge has drifted obliquely into the subduction zone. The more northerly part of the ridge intersected the more northerly part of the trench, and then the point of intersection drifted southerly along the trench axis.

The northern part of the Tonga subduction zone is typical of such zones. Starting in the east, the zone has a flexed arch, a trench, a ridge capped by atolls, an active volcanic arc, a back-arc basin, and a zone of earthquakes dipping west from the trench. This northern region thus shows no signs of disturbance by a buoyant volcanic ridge.

The southern part of the Tonga subduction zone is where the Louisville ridge apparently has been subducted. From east to west, a profile shows no flexed arch, a trench, a ridge with an uplifted atoll, no active volcanoes, no back-arc basin, and a lack of earthquakes at shallow depths. The demonstrable absence of a flexed arch in old subducting lithosphere appears to be unique to this region. Everywhere else, there is either an arch or the bathymetry is obscured by a ridge or plateau intersecting a trench at a high angle. The Louisville ridge and Tonga Trench intersect at an acute angle, so the area where flexing should occur is exposed. The reason for the existence of an arch elsewhere seems clear enough and can be simulated in a laboratory. If the plunging oceanic lithosphere is pulled down by gravity, the adjacent, not-yet-plunging oceanic lithosphere is flexed upward in an arch. If there is no arch off the Tonga Trench, the adjacent "plunging" lithosphere must not be plunging. If gravity is not pulling it down, it must be relatively buoyant, even though it is subducted, so it is drifting horizontally under the Tonga region, and there is no reason for the development of a compensating arch. The cause of the other features characteristic of this region of the Tonga subduction zone is much less certain. The uplift of 320-meter Ewa Island, a former atoll, presumably reflects the buoyancy of the subducted Pacific plate. The absence of active volcanoes is characteristic of those parts of subduction zones where volcanic ridges are subducted obliquely, but just why is speculative. Normally, andesitic magma is generated at a certain depth by heating of the subducted oceanic crust and sediment. Presumably, the horizontally drifting subducted Pacific plate is not deep enough to generate such magma, so no lava emerges at the sea floor to build an island arc. Likewise, it appears that the gap in shallow earthquakes indicates that such earthquakes are somehow related to the process of plunging instead of mere subduction. Much remains to be learned, but it is evident that almost every major tectonic feature of subduction zones can be altered by a relatively minor decrease in density of the subducting plate. Islands are hard to digest.

ACCRETED TERRANES

In the past decade, geologists have found that several continental regions consist of accreted fragments derived from other continents and oceanic crust. A notably well-studied region is western North America. The fragments, called *terranes,* can be identified as such by their differing lithology, fossil assemblages, age, and the orientation of paleomagnetic indicators. Among the terranes are some consisting of volcanic rocks, which once were seamounts, oceanic islands, ridges, and plateaus.

Oceanic volcanoes that have been obducted or subducted and later exposed by erosion can be recognized in different ways. Perhaps the most interesting is the recognition of Paleozoic volcanic rocks in Norway as derived from oceanic volcanoes. Associated with the rocks are fossil species found nowhere else. It appears that they were endemic when alive, and this implies isolation and evolution such as commonly observed on existing oceanic islands.

More commonly, oceanic volcanoes accreted recently in geological time can be recognized as topographic bulges and bumps on the inner walls of trenches. Older oceanic volcanoes are identifiable by the distinctive composition of their igneous rocks. Several have been found by drilling in the accreted terranes next to modern trenches or by surface sampling in island arcs. Ancient oceanic volcanoes—ranging from Cenozoic to Paleozoic age—have been incorporated in many places into continents. For example, large volumes of volcanic rock in the state of Washington may have originated as oceanic volcanoes. Likewise, Ordovician rocks in Newfoundland have the composition of islands or seamounts and are voluminous enough to have been an oceanic ridge. Moreover, associated carbon-

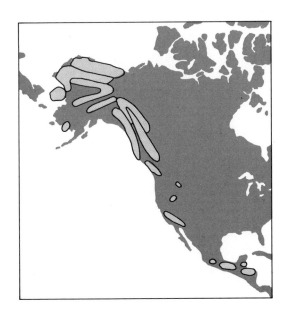

Much of western North America is made up of terranes, accreted fragments of other continents and oceanic crust.

ate rocks (not reef faunas) indicate that the ancient submarine peaks were shallow.

Inevitably, if continents and oceanic volcanoes coexist and drift about, their paths will intersect. Apparently many of the volcanoes become incorporated into continents instead of being subducted into the mantle with the oceanic lithosphere. How much of the volume of continents could have been derived by accretion of oceanic volcanism? Even the most speculative estimate depends on whether the rate of volcanism was constant and whether archipelagic aprons are included. The volume of continental crust is about 7×10^9 km^3, and the volume of oceanic volcanic edifices and roots is roughly 10^7 km^3. The average age of the ocean floor is about 5×10^7 years, and the age of continents is about 4×10^9 years. Taking oceanic volcanism as constant, a maximum of about 10 percent of the continental volume could be former oceanic volcanoes. If archipelagic aprons are included, the fraction rises to 30 percent. However, obduction as opposed to subduction seems to be favored by rocks of relatively low density. The midplate, abyssal, flood basalts of archipelagic aprons probably are at least as dense as the normal volcanic rocks generated at spreading centers, so they would tend to be subducted rather than accreted.

Even if oceanic volcanoes have added significantly to the volume of continents, little of their original rocks remains. Continental rocks are constantly being recycled by erosion, so the once distinctive minerals that formed a very ancient island have long since been weathered and dispersed. The submarine volcanic platforms of islands can survive the dramatic events of subduction, but nothing can long resist erosion.

10 *Island Life*
From a Geological Viewpoint

In the middle of the nineteenth century, the era of Charles Darwin and Alfred Wallace, it was widely appreciated that islands are evolutionary laboratories. By 1880, when Wallace published *Island Life,* he thought that the principal problems had been solved: the isolated islands were populated randomly by waifs, and the strange life forms on the islands were a result of local evolution since the waifs landed. Hardly anyone would dispute his judgment now, although a century of intense controversy has intervened. Nonetheless, the study of island life continues just because of the laboratory-like circumstances and the small range of a very limited number of environmental variables. Moreover, concepts of the general geological history of island life have recently been rejuvenated by the discoveries of plate tectonics. The islands move horizontally and vertically and thereby grossly modify the environment on and around them. Life forms too must evolve, migrate, or become extinct as the land changes under them.

How this happens depends much on the nature of the organisms that manage to reach islands. Open expanses of ocean are impenetrable barriers to some land organisms, permeable filters to many waifs from land, and highways to marine organisms. Moreover, the width of the open ocean directly determines the permeability of the filter, or, to put it differently, the probability that a given type of organism will reach an island. These effects of distance are admirably demonstrated by the vertebrates of the Melanesian region of the southwestern Pacific. Most of the islands from

A monk seal, found only in the Hawaiian Islands.

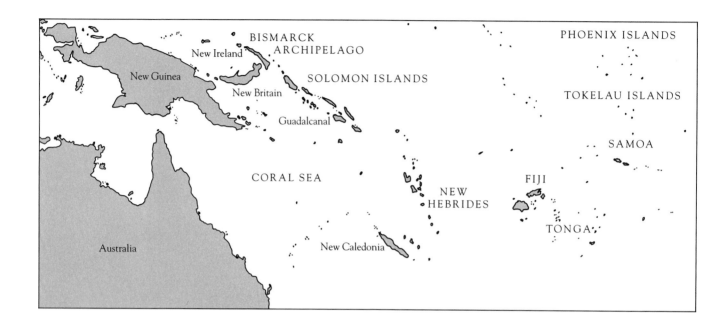

The islands of Melanesia are stepping stones for the spread of plants and animals from Australia and Asia into the heart of the Pacific.

New Guinea to Tonga are not oceanic in a geological sense, but they are relatively small and rise from deep water, so the filtering mechanism resembles that of atolls and volcanic islands. The source region for the waifs was the great continental land masses of Australia and southeast Asia. Starting from the source region, freshwater fishes and monotremes such as the duck-billed platypus have not migrated beyond New Guinea. The marsupial bandicoots and wallabys extend only to the large islands of New Britain and New Ireland just east of New Guinea. The last terrestrial mammals and, probably, frogs are filtered out in the next group of islands, the terrible Solomons of World War II. Snakes and lizards are found farther east, but the last snakes are in Fiji and the last lizards in Tonga. Thus, in a distance of 4000 km from continental New Guinea, despite a wealth of island stepping stones, all continental vertebrates except birds (and a few lizards) are filtered out. North and east of Tonga, beyond the filters, is the great expanse of the Pacific Ocean.

There are other pathways and systems of filters between other continents and the islands of ocean basins, but the same general rules apply, as shown in the table on this page giving the extreme oceanic distances successfully crossed by the main groups of land and fresh water vertebrates. (A successful crossing means one leading to a new population on an island.) It should be remembered in examining the table that the number of

Width of the widest oceanic gap crossed by various groups of animals

Group	Width of gap
Freshwater fishes	5 km
Elephants and other large mammals	50 km
Small mammals except rodents	400 km
Rodents (to Galapagos)	1100 km
Amphibians	1000 km
Land tortoises (to Galapagos)	1100 km
Snakes (to Galapagos)	1100 km
Lizards	1800 km
Bats (America to Hawaii)	3600 km
Land birds	3600 km

waifs that had a chance to make a crossing probably was very large during geologic time.

Certainly from our point of view, the most important and far-reaching biological event in the geological history of islands occurred within the last thousand years when a mammal, man, at last burst among the oceanic islands. Apparently, nothing like man, with his entourage of rats, pigs, chickens, and goats, had gone before. Thus, nature's small island-laboratories show one more thing—how man interacts with his environment.

EVOLUTION

Evolution, small and relatively simple on islands, has been grand and complex on continents. Despite the complexity of the geological record of life, generations of paleontologists have succeeded in deciphering it by their labors in field and laboratory. In its essential elements, the message of the fossils is simple, clear, and overwhelmingly consistent. The earth formed 4.55 billion years ago. From time to time since, the sedimentary rocks that are now at the surface of the earth were deposited—locally, regionally or globally—on the rocks below in a sequence like a layer cake. Parts of the sequence have been deformed and melted, but identifiable pieces more than 3 billion years old still remain. At many places, the sedimentary layers can be dated by the decay of radioactive minerals in interlayered igneous rocks. The sedimentary layers contain enormous numbers of fossil remains of organisms that lived during the past 3 billion years. Almost all of the fossil species are now extinct. Just how or why they became extinct is an exciting and generally controversial question, but the fact of extinction is certain. Indeed, the fossil record is so extensive that the whole life history of long-dead groups of animals and plants is known. The most famous of these were the dinosaurs. They first evolved roughly 225 million years ago and became extinct about 65 million years ago. In 160 million years, they spread to most and probably all of the present continents. The earliest dinosaurs evolved by *adaptive radiation*—they became herbivores and carnivores able to live and feed in the ocean, rivers, lakes, swamps, and, in many climatic zones, on land. Some were as small as dogs, but others were the largest animals that ever lived on land.

In less dramatic form, countless other animals and plants evolved and radiated to fill ecological niches. Life apparently originated in or at the edge of the sea, then it spread to the land and finally to the air. The history of flight illustrates another important fact about life. Some dinosaurs learned to fly; so did birds, insects and mammals including both bats and, in their way, men. Thus, similar forms and functions may arise from very different ancestors.

Much remains to be learned, or agreed upon, about the cause of evolution, but several important factors have been identified. One of these is the role of continental drift and plate tectonics. Imagine a great continent such as a combined Africa and South America. The plants and animals would differ from place to place because of climatic zones. However the members of a species could mingle within large areas and thereby maintain an interbreeding population. Next, imagine that the continent splits in half, and an Atlantic Ocean gradually opens between the pieces. Evolution gradually changes the plants and animals on each of the new continents, and, after geological ages, they are as different as the animals of Africa and South America. A different effect occurs when two continents collide in a subduction zone. The animals, to limit the analysis, of each continent may have radiated to fill all the ecological niches. However, quite different animals may inhabit the lakes or high grassy plains of the two continents. Thus, competition on a grand scale may ensue when one population encounters another.

The duration of the competition need not be lengthy if one group of animals can simply kill the others instead of surpassing them by less violent means of natural selection. The fate of the large and distinctive mammals of late Pleistocene time clearly was some sort of rapid extinction, but

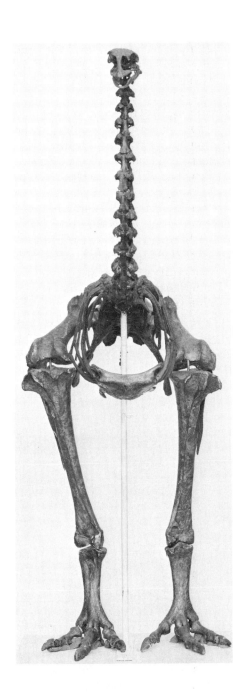

The skeleton of a moa.

how? Alfred Wallace first thought that such animals as the imperial elephant, sabre-toothed "tiger," and giant ground sloth probably were killed as a consequence of rapid changes in climate. By 1911, he had realized that a far more potent killer than glacial ice or desert heat had been loosed in the late Pleistocene.

Man himself probably was the mass exterminator. The issue still is in dispute, but the circumstantial evidence for the murder is very strong. The major extinctions of large mammals were surprisingly recent and they coincided rather closely to the migration of mankind. By 40,000 years ago, the large Pleistocene mammals were largely gone from the ancestral home of man in Africa, from southern Asia, and from Europe. The cave drawings of France and Spain show many species that would soon be extinct. Rapid extinctions occurred in the period from 8,000 to 13,000 years ago, as humans arrived in the Americas and Australia. There were later extinctions on the outlying islands of Madagascar and New Zealand, including the giant flightless birds, moas, who were hunted by Maoris and did not survive until Europeans arrived.

SPECIAL FEATURES OF ISLAND EVOLUTION

Islands are small and isolated, but they are populated by airborne seeds, birds, and, possibly, marine mammals almost as soon as the active volcanoes grow above sea level. We call an island "uninhabited" if men do not live on it, but one of the joys of being an explorer is to stroll or swim among the teeming life forms of such an island. So-called "desert islands" abound with rookeries with hundreds of thousands of birds or thousands of sea elephants, and the ground is covered with curious plants. The special characteristics of island life derive chiefly from three factors: filtering out of most continental species by the rigors of long-range dispersal, the isolation of the successful immigrants, and the availability of a large range of ecological niches that are not filled by the usual continental occupants.

One of the principal problems for a species after long-range dispersal is that the number of migrants in any episode may be very small and there may be only one episode of migration. Thus a single breeding pair of beetles might be blown to an island and have the potential to thrive. However, the chances that they would successfully establish the species would be very small. Even if scores of a species arrived in a single storm they would be able to draw on only a small pool of genetic material. Inbreeding would prevail, and the species would be relatively vulnerable to such factors as environmental change or a new disease. If there were repeated migrations of the same species, the genetic material would be constantly enriched and the vulnerability of the islanders would scarcely be greater than that of the continental source population. However, the uniqueness or rarity of most migratory events can be seen in the nature of island populations.

Consider first the biota of the shoreline as opposed to the interior of an island. The pandanus, coconut, and casuarina trees of the insular shorelines are the same all over the tropical Pacific. Newly arrived coconuts are usually conspicuous on the beaches and outer reefs of uninhabited islands, so it is obvious that the genetic material of these plants is constantly renewed and dispersed. A close look at the molluscs and corals of tropical islands reveals that they too are remarkably similar from island to island. They have larval forms that drift frequently around the open ocean and reestablish themselves on islands. Nevertheless, the coastal faunas are not exactly the same. Adaptive radiation and evolution affect some of the marine life of islands. Thus, 34 percent of the shore fishes of Hawaii are "native" or *endemic* as are 50 percent of the echinoderms of St. Helena. Pelagic fishes such as tuna may travel frequently from one region to another, but many shore dwellers apparently are descended from waifs.

The plants and animals of the interior regions tend to be even more distinctive than those of the shore. Endemic species abound on different islands or even different areas of one island. The general types or classes of organisms are the same because they are the ones that are capable of

Coral spawning.

migrating for long distances. Having arrived, however, they obviously have been changed by evolution into new species.

Island Size

The factors that influence the origin of endemic species in general are many. On oceanic islands, area appears to be important. For example, the number of species of amphibians and reptiles on certain islands in the West Indies increases systematically with the area of the islands. About 80 species live on an island of 70,000 km^2, 40 live on one of 7,000 km^2, and so on down. Many of these species probably migrated from nearby continents and continental islands but the same general relation may be expected on an isolated oceanic island. If only a single species succeeds in populating a large, mountainous island, it can be expected to evolve into many species by adaptive radiation until a whole range of environments is occupied.

Rate of Evolution

Atolls are old and high islands are young, so it might be expected that endemic species would be more numerous on the former. The reverse is true, because atolls constantly receive migrants and because endemic species can arise very rapidly in the interiors of high islands or wherever competition is lacking. Rates of evolution are known very accurately where species were introduced to an island by humans. Five new species of banana moths have evolved in Hawaii since the Polynesians introduced the banana to Hawaii only about 1000 years ago. Similar prodigies seem to have occurred in the Hawaiian archipelago in the geologically recent past. There are, for example, about 400 endemic species of fruit flies; and land and freshwater birds have evolved beyond the species level to endemic genera and even one endemic family. The birds include endemic geese, ducks, coots, owls, hawks, crows, thrushes, and flycatchers, which, having evolved, have never migrated elsewhere.

Flightless Birds

A surprising number of endemic species of birds on islands are, or were, flightless. In the absence of mammals, reptiles, and other predators, some birds evolved greatly to occupy the environmental niches normally filled by rabbits or sheep. About two dozen species of ostrich-like moas lived in New Zealand, and many were two or three meters tall. *Aepyornis*, a flightless bird of Madagascar, was even larger. All these remarkable giant birds on islands are dead; when they were finally confronted with a predator,

Left: Dodos, extinct flightless birds, evolved on small islands in the Indian Ocean from birds that had flown or been blown there. *Right:* A flightless cormorant (not extinct) endemic to the Galapagos Islands.

they were unable to survive as did the emu and the ostrich, on the continents.

It was a flightless insular bird that gave us a figure of speech meaning extinct—"dead as a dodo." The dodo was a singularly ugly bird that was immortalized by Sir John Tenniel in one of his illustrations for *Alice in Wonderland.* It was the dodo who decided that everyone had won the caucus race. The dodos included three species of flightless birds on the islands of Mauritius, Reunion, and Rodriguez, in the Indian Ocean. They were descended from a ground-feeding variety of pigeon and presumably evolved separately after flying to the islands. They grew up to a meter tall and waddled about functioning as terrestrial herbivores, like avian sheep. One was exhibited live in London in 1638, and it, and another in Holland, were painted by many artists before Tenniel. The birds were relatively helpless, and they were exterminated by 1700. Man himself was not interested in killing dodos because they were notoriously foul smelling and inedible and the skins had no value. Apparently, hogs and dogs ate the young birds and eggs. The adults had a mean peck, but it was not enough.

Relicts

One concept that has changed greatly in recent years is the belief that the faunas and, particularly, the floras of some islands are ancient relicts.

Darwin observed that there are no Paleozoic or Mesozoic fossils on oceanic islands, and none have since been found—except for Cretaceous ones on guyots. Thus, there was never any question of living relict plants and animals of such geological antiquity. However, botanists, including Darwin's crony Sir Joseph Hooker, insisted that the floras of some islands were descended from Tertiary rather than recent continental plants. It was not that plants on islands could not have evolved enough in only a few million years. It was that they had common ancestors who had died out on the continents in Tertiary time (65 Ma to 2 Ma ago).

Now, potassium-argon dating has shown that almost all the high islands are only a few million years old. Consequently, the significance of the botanical evidence has come into question. Indeed, the temporal and stratigraphic boundaries of the subdivisions of Tertiary time have been changed by geologists repeatedly during the last century. Thus, the mere meaning of "Tertiary flora" or "Miocene flora" as applied to a given island may require expert opinion. The consensus at present appears to oppose the existence of spectacularly old insular relicts, but a high degree of endemism still remains difficult to explain. On the shores of Easter and Sala y Gomez islands, for example, some 42 percent of the species of marine molluscs are endemic. The islands are very small and only 2 Ma to 2.5 Ma old. It appears that the necessarily rapid evolution in such limited circumstances should tend to produce more extinctions and fewer new species. The fauna cannot have been derived from the shores of nearby islands, because none are closer than 2000 km. Thus, W. A. Newman and B. A. Foster reason that the molluscs are at least partially relict from the former shorelines of the abundant guyots in the region.

POLYNESIAN CULTURE

In his great history of cultural evolution, Arnold Toynbee, listed three civilizations that were arrested in development by the environment. The Polynesians, Eskimos, and Nomads respectively conquered the tropic sea, the polar snow, and the desert by what Toynbee regarded as a *tour de force.* The Polynesians, for example, by audacious voyages reached the extreme limits of the powers of their culture. They could do no more and remained in a precarious equilibrium with their environment for long ages (so he thought), until the Europeans arrived. There is much to be said for these views except, perhaps, for the idea that the equilibrium was precarious. It may, in fact, have been quite stable. The instability lay in the potential for intrusion by an alien culture that would not be content with equilibrium or able to reach it with the island environment.

Huahine, in the Society Islands, as painted by William
Hodges, who voyaged with Cook and saw Polynesian society relatively unaffected by European contact.

Consider, for example, the relations between Polynesians and atolls.
An atoll has hardly any land, so there is no way for a large human population to accumulate. An atoll has a long shore perimeter for its area, and it
has an enormous lagoon. Hence there is a very large source of easily ac-
quired protein. A few vegetables and fruits are easy to grow and, in the
right latitudes, rainwater is abundant. An atoll is an environment almost
ideally designed so man cannot get out of equilibrium with it. It might
have taken a catastrophe like an ice age or a giant wave from a large
meteorite impact in the ocean nearby to dislodge Polynesians from an
atoll. They certainly knew how to survive hurricanes and their high
waves.

Polynesian culture on high islands may also have been stable with
regard to the environment. The very characteristics that, by some classifi-
cations, would identify the culture as arrested would make for stability.
There was no metal for weapons or clay for bricks. Consequently there was
no incentive to burn the native forests in kilns or forges as in the early
civilizations of India and elsewhere. Moreover, the island interiors were
very steep for farming, so there was no incentive to clear the forests for
that. The stability of the rivers and forests was not due to any equilibrium

between large herbivores and carnivores. Thus, there was no opportunity for the Polynesians to upset the equilibrium by killing the carnivores. By a mercy, the Polynesians did not bring goats or rabbits, so the forest growth was not destroyed by feral animals. The end result of all these facets of Polynesian culture and its environment was that the forests were preserved and with them the soil, watersheds, water supply, and the coastal food supply that depended on clear water. Moreover, given isolation, the equilibrium with the environment was not apt to change significantly. Certainly no need for kilns or forges would have arisen on oceanic islands. Of course, continental islands are different. Once the Polynesians arrived in New Zealand and had the giant moas to exterminate, they exterminated them.

INDUSTRIAL CIVILIZATION

The unarrested industrial civilizations of Europe, America, and Japan have proved capable of destroying whole islands and, without even trying, their biota. The stable relations between mankind and atolls only persisted as long as the users of an atoll were also the permanent occupants. Successively the Germans (in WWI), the Japanese, and the Americans conceived of atolls as fleet anchorages and fortified them as advanced bases. Then, in 1941, the Japanese launched the first modern battles on atolls and captured Wake from the United States but were disastrously defeated at sea west of the atoll of Midway. The American counterattack in the central Pacific began with a naive assault on Tarawa atoll in the Gilbert Islands. The inexperienced naval commanders believed that there would be scarcely any organized survivors on the tiny islands of the atoll after massive shelling and bombing. It was part of my job at that time to assess such damage, and we all tended to forget the trivial effects of endless shelling of the trenches in World War I. An atoll has no relief and no cover. The Marines assaulted across the reef into still intact fortifications, and a thousand died in four days before the last of 4700 Japanese defenders was killed.

The amphibious landings at Eniwetok (now Enewetak) and Kwajalein went a little better for the attackers, but each of these victories required the destruction of every last small bunker and gun emplacement. The aftermath was the utter destruction of the surface of one or more islands per atoll. However, the atolls survived. Rain fell and water percolated down to the Ghyben-Hertzberg layer; solution and precipitation of lime-

An American fighter bomber over Wake Island
in World War II.

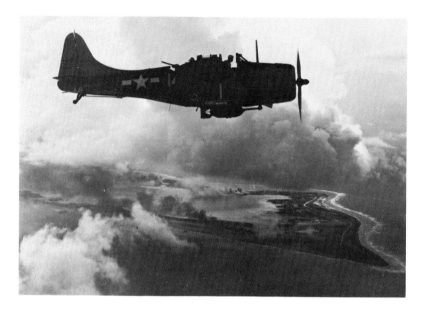

stone began to cement the exploded fragments together; and plants and
animals migrated from the undamaged islands to repopulate the barren
landscape. Except for the danger of unexploded ordnance, the atolls were
essentially restored within half a dozen years. The reef flats were not as
attractive because they were imbedded with bits of landing craft, bombs,
and beer cans, but these were rusting away. Certainly the conventional
weapons of World War II were unable to end the long lives of atolls.

With the dawn of the nuclear age, industrial man conceived of a new
use for atolls. After a few residents were persuaded to leave, an atoll could
be an ideal site for nuclear tests. It was isolated, provided a good anchorage, and, except to the former residents, was not much good for anything
else. So Bikini, in the Marshalls, became famous as the site of a shallow-
water nuclear test in the lagoon. Enewetak was used a few years later to
test the first hydrogen bomb. It vaporized the test island. Within a few
years, I was on each of these atolls, which looked little the worse for wear.
However, the longer-term effects of radiation on the atolls and on the
former human residents who sought to return remains to be determined.
Meanwhile, Great Britain tested its first hydrogen bomb on Christmas
Island in the Indian Ocean. This former atoll, uplifted on the outer arch of
the Indonesian subduction zone, had already been defaced by phosphate
mining. The French, not to be outdone, established a (still) permanent
test site at Mururoa atoll, in the Tuamotus.

Atomic bomb test, Bikini, 1946.

High Islands

The history of Saint Helena highlights the differences between European and Polynesian cultures. When the Portuguese discovered the island, in 1501, it was so luxuriantly forested that trees fought for space by hanging over the high cliffs. In 1513, the Portuguese introduced goats, which, seventy years later, numbered in the thousands. The goats ate the seedlings in the forest. In 1651, the British East India Company took over the island and felled the redwood and ebony trees, because their bark was useful for tanning. In 1709, ebony was freely burned to produce lime for fortifications. In that same year, the governor of the island informed the Directors of the East India Company that the forests had largely been killed and that the bare ground was being eroded away by the same abundant rainfall that had supported the vegetation. He recommended killing the goats, but the Directors decided that they were worth more than the ebony. All this history was recorded in 1880 by Alfred Wallace, who added that, by 1810, the forests were gone and the British Government had to spend £2729/15/8d in one year to bring in firewood. Plants were then imported haphazardly from several continents to convert the barren wastelands into a prison suitable for an emperor.

Wallace was also outraged about the wanton slaughter of endemic species on other high islands, but, in general, the islands themselves have

Goats run rampant in this acquatint from
Views of St. Helena, 1815.

fared better than Saint Helena. Perhaps the principal environmental changes have been a consequence of farming for export. The soils are very fertile because of the nutrients introduced in both the primary and miderosional phases of volcanism. Moreover, the orographic patterns of rainfall produce cloudy regions of abundant water almost side by side with sunny regions ideal for rapid plant growth. With aqueducts for irrigation or wells to tap groundwater, intensive agriculture became possible. Sugar cane, pineapple, and other tropical fruits that could be canned were culti-vated for export on such remote oceanic islands as Hawaii, Raratonga, and Mauritius. Most islands, however, were merely farmed enough to support the local population and to supply fresh produce to ships.

    After all the islands were explored, ships tended to visit only a few that had good harbors, adequate water and supplies, and a friendly popula-tion to "refresh the crew," in the eighteenth-century phrase. Most islands in high latitudes were ringed with great cliffs and had little appeal. Even in the tropical Pacific, the young islands generally lacked good harbors. In contrast, old atolls were nothing but harbors. Clearly, the ideal combina-tion would be found in barrier-reef islands of intermediate age. By far the most important port on an oceanic island was Pearl Harbor on Oahu. The barrier reef is poorly developed on the high inactive Hawaiian islands, but one exists to protect the great naval base. The harbor has several arms, or

A coral-encrusted tank at the bottom of Truk lagoon.

"lochs," which mark the course of drowned river valleys. The relatively flat coast is ideal for docks and storage. Nothing comparable exists for 4000 km in any direction.

The great fleet base for the Japanese navy was far to the west among the old atolls of the Caroline Islands. It was Truk, a barrier reef surrounding an enormous lagoon sprinkled with the erosional remnants of the peaks of an ancient volcanic island. As it happened, neither Pearl Harbor nor Truk was able to defend itself against the assault of dive and torpedo bombers from enemy aircraft carriers. Both harbors are strewn with the wrecks of warships from World War II. The major difference is that the hulls of the winning side are national shrines and those of the losing side are distant attractions for diving tourists.

The third famous harbor of the central Pacific is Papeete, and it, too, is in the lagoon between a barrier reef and a high island—Tahiti. Papeete was far to the south of the great naval war, so, as late as the 1950s, it remained a romantic, storied haven in the South Pacific, a port and an island pictured by Gauguin instead of in photographic postcards. However, Tahiti had characteristics that would encourage change: a harbor that could be expanded behind the natural breakwater of the reef, an area of broad reef and shallow lagoon suitable for an international jetport, a colonial government, and suitable facilities and space to build a staging base for the nuclear tests at Mururoa, to the southeast. So Papeete has most of the aspects of the French Riviera, including four kinds of real champagne in any ordinary grocery market. There is an evocation of earlier bomb tests in the bathing costume that the French made famous as the bikini. Of earlier times, little remains. The beautiful Gauguin gallery contains not a single original work by the artist, and the model Tahitian village was built for a remake of the motion picture *Mutiny on the Bounty*.

LIFE ON A DOWN ESCALATOR

The geological history of islands is a basic factor affecting life on them. The principal facts regarding high islands are that almost all sink below the sea in 5 Ma to 10 Ma. During that time they drift as much as 1000 km. Moreover, most are created by hot spots that persist for 100 Ma and produce islands only 10 km to 100 km apart for much of that time. Thus, geology and plate tectonics affect life on islands mainly by changing the terrain by erosion, subsidence, and reef building; by drifting islands across the edges of climatic zones; and by providing sequential stepping stones whereby plant and animal life can evolve for the age of a hot spot instead of just the age of an island.

Erosion and Subsidence

Young oceanic islands are active volcanoes, commonly ringed by cliffs that grow ever higher as volcanism wanes. For most life forms, young islands are most inhospitable, and yet plants on Hawaii and in the Galapagos have adapted themselves to a life on raw, still-warm lava flows. Thus, even the most youthful stage of geological evolution is accompanied by life, despite the virtual absence of soil. Natural volcanic ash with rainfall makes an effective planter's mix for seedlings.

A peculiarity of small, young islands is the dichotomy in elevations and slopes. The waves carve a broadening coastal terrace, and the uneroded core of the island retains its gentle slopes. Meanwhile, the boundary between the nearly horizontal slopes is an almost vertical cliff hundreds of meters high. Life must adjust to the changing landscape. It may be that the lush forests of St. Helena overlapped the cliffs not because of herbaceous exuberance but because the ground was being cut from under them. The landscape usually evolves rapidly in volcanic islands less than 0.5 Ma old. The central peaks are gutted to great amphitheaters by torrential new

Cactus grows on fresh lava on Fernandina Island in the Galapagos.

rivers. Planeze surfaces remain as havens for the life of the former central plateaus, but the ground changes rapidly. Perhaps it is no wonder that adaptive radiation is also rapid.

In a few million years, a mere instant in geological time, most of the islands in high latitudes are gone. Whatever evolution may have taken place to adjust to erosion and wave truncation is over, unless somehow the evolved species can escape to a younger island. In the tropics, the problem of survival is not simple drowning in water but drowning in coral at about the same rate. Perhaps a few plants or animals of the shoreline might evolve to meet the challenge. However, the reef environment is already fully occupied by forms that generally migrate freely; at every stage, forms newly evolved on the island are in competition with older established ones from elsewhere. The characteristics of atoll plants and animals indicate that the establishment has been winning. Thus, in the tropics as well as the temperate zones, the principal path for continuing insular evolution is island hopping.

Drifting through Climatic Zones

Few high islands last long enough to drift from one climatic zone to another. However, if a hot spot is in just the right place, the effects of drift can be significant. The most notable example in recent geological history is the Hawaiian hot spot. Paleomagnetic studies indicate that during Tertiary time the hot spot has stayed at about the same latitude—which is at the northern edge of the tropics. Moreover, the Pacific plate has drifted northwesterly, thereby carrying the new islands out of the tropics even before they are truncated by erosion. As a consequence, the Hawaiian Islands do not form a perfect sequence of increasing submergence like the Society Islands, which are drifting into the tropics. Notably missing in Hawaii are the equivalents of the magnificent barrier-reef islands of Bora Bora and Huahine. Instead, the Hawaiian Islands present a sort of statistically disturbed sequence of tropical submergence. High islands, pinnacles, aborted barrier reefs, banks, atolls, and guyots are somewhat mixed, although the average morphology is that of a subsidence sequence. We have observed elsewhere that the intermixing of morphologies is a natural consequence of midplate volcanism. In Hawaii, however, the cause probably is different, because the dated age sequence favors a single midplate hot spot. What has varied is the climate at the edge of the tropics and with it, apparently, the ability of vigorous reef growth to become established. Judging by their morphology, most of the northern Hawaiian reefs are a frail lot, quite unable to protect a volcano from wave erosion or contain the products of stream erosion in a vast lagoon. Probably most of the great submarine banks north of Kauai are truncated volcanic islands only thinly covered with coral. Midway, however, has normal structure for an atoll. It has mountainous, untruncated basement topography under a coral capping hundreds of meters thick. At Midway, at least, the initial reef was healthy enough to survive the long drift out of the tropics.

At present, the world has no good sequence of islands drifting from coral-free seas into those with coral. The nearest thing is in the Austral Islands, where the MacDonald hot spot is in just about the right position. Indeed, Rapa, which is a 5-Ma product of this hot spot, has a barrier reef of sorts. The reef is largely submerged, which is why a Scripps ship grounded on it in the 1950s. However, two factors oppose the rescue of drowning Austral Islands by living reefs. For one, the hot spot is not very voluminous, so most of its islands are long since truncated and submerged before reaching the tropics. The other reason is that plate drift in the Austral region is more to the west than to the north. If the drift ever changes to due north, even the small volcanic islands probably would survive into the warm waters of the tropics and become atolls.

Whether any group of high islands develops a submergence sequence grading into atolls depends on more than latitude, and the other factors may influence drifting island life. Most significant, probably, is the presence of upwelling and cold currents. The western shores of oceans are notably richer in corals than the eastern ones, and this is because of the general circulation of the ocean. Cold currents flow from the polar seas along eastern shores and from tropic seas along western ones. Thus, the great barrier reefs of Australia and Belize are to the west, and the Galapagos, with their tropical penguins and no reefs, are in the east. The boundaries between cold and hot currents can be quite narrow zones across which a high island could drift. The coralless Juan Fernandez Islands, off Chile, are in just such a zone on the eastward-drifting Nazca plate. However, they are already in a temperate, untropical climate, as Alexander Selkirk, the real-life Robinson Crusoe, learned. Under slightly different circumstances, established atolls could very easily drift east or west into waters too cool for coral.

Upwelling of deep ocean water is a consequence of the rotation of the earth interacting with ocean currents. In order to conserve angular momentum, surface currents moving toward the equator off California veer to the right. Off Chile, equator-bound currents veer left. Off either California or Chile, the veering is away from the coast—which leaves a gap that is filled by upwelling water. The deep water is cold and extremely fertile because of high concentrations of nutrients from sediment. Thus, off a coast with upwelling, one has fog and commercial fisheries such as the California sardine and Chilean anchoveta. These circumstances raise another possibility for drifting island life. Phosphate-rich islands are created by birds depositing enormous quantities of guano; for them, an even more enormous local source of food is essential. In slightly different geography, the Juan Fernandez Islands would be in just the right position to drift east into the edge of the zone of coastal upwelling before they subsided completely below sea level. Thus, similar islands may have drifted from the relatively sterile waters far offshore into fertile waters with upwelling. Thereafter the scenarios become rather wild because of the effects of overriding outer arches, subduction, buoyancy that balances subduction, and so on. Nonetheless, mention should be made of another possible kind of drowned ancient island. An atoll descending into a trench may be coral from top to bottom; this is possible because living coral can easily grow upward faster than the base of a volcano subsides by subduction. But guano also accumulates much faster than subsidence; thus, phosphatized guano may possibly be a kilometer or more in thickness on an island in the right circumstances—a subduction zone on the eastern boundary of an ocean, such as off Chile and Peru now or California as it used to be.

Sequential Island Hopping

One of the early ideas for populating islands in the middle of an ocean was by hopping short distances along a chain of islands stretching to a continent. The fact that no such chain existed—to Hawaii, for example—was not viewed as a major flaw in the argument. The mountains under the sea were unsurveyed before 1950, and therefore there might be countless drowned islands. Charles Darwin, who believed in dispersion by waifs, in fact had predicted that oceanic islands had been drowned. The advocates of dispersal by island hopping were greatly strengthened in their beliefs by the discovery and mapping of chains of guyots in the 1950s. E. L. Hamilton and I lent support to the concept by our dredging and mapping of guyots. Numerous drowned stepping stones actually existed from Australia and Asia to Hawaii, and there were a few toward North and even South America. The steps were 60 Ma to 100 Ma old, but at the time that provided no problem. No one could prove that the base platforms of the high islands of the Pacific were not equally old. Within two decades, the original argument collapsed, because isotopic dating and the hot-spot hypothesis showed that high islands are young from the sea floor to the mountain peaks. The stepping stones had been drowned long before the high islands formed.

Gardner Pinnacle, rising from a submarine bank about 1000 kilometers northwest of Kauai, is the oldest high island in the Hawaiian chain. The pinnacle and other submarine banks nearby are all that remains of an archipelago of large high islands that teemed with life fifteen million years ago where the major Hawaiian Islands are now.

Scientific hypotheses are hardy, and many survive the death of the original supporting evidence. So it was with stepping-stone dispersal. If it could not work all at once at one time, perhaps it could work through time as a steady-state phenomenon. That is the present status of island hopping. The very concept of sequential development of islands that drift away from a hot spot seems to have the corollary that plants and animals can migrate counter to island drift. If the botanists believe that the Hawaiian plants are much older than the islands, why not? The plants can have been migrating along the archipelago for all of Tertiary time. Not a great long chain of an archipelago reaching like the Emperor guyots all the way to Asia. Nothing like that. Just a short line of high islands like the present ones—born at Hawaii and eroded away not far north of Kauai. Thus, in the present hypothesis, the plants and, conceivably, animals are essentially migrating up a down escalator. It is a peculiar escalator, with steps separated by gaps, and the ability to negotiate the gaps is one of the most essential characteristics for long-term survival.

Nothing in our present understanding of geology prevents evolution for a hundred million years on the islands of a hot-spot track. Whether survival of the descendants of a few waifs is likely under such circumstances is another matter. However, evolution and survival have certainly taken place on small drifting continental fragments such as the Seychelles, so perhaps they have also on oceanic plateaus or ancient Icelands, long-lived enough to drift across an ocean without sinking beneath the waves. If so, obduction may have exposed continental faunas and floras to many a return invasion by the cast-up waifs of the sea.

Sources of Illustrations

Drawings by Tom Cardamone Associates

FACING PAGE 1
ET Archive, Ltd.

PAGE 2
Erwin Christian

PAGE 8
India Office Library and Records

PAGE 9
Douglas Faulkner

PAGE 11
Elliot Varner Smith

PAGE 13
Douglas Faulkner

PAGE 16 (top)
Bernice P. Bishop Museum

PAGE 16 (margin)
Collection of Don Severson

PAGE 17
The British Library

PAGE 18
From the Collection at Castle Howard, York, England

PAGE 20
After P. S. Bellwood, "The Peopling of the Pacific," *Scientific American*, November 1980

PAGE 21
Ric Ergenbright

PAGE 23
Jeff Foott/Tom Stack & Associates

PAGE 26
Gudmundur E. Sigvaldason

PAGE 30
F. J. Vine, "Magnetic Anomalies Associated with Mid-Ocean Ridges," in *The History of the Earth's Crust*, edited by Robert A. Phinney, Princeton University Press, 1968

PAGE 32
H. W. Menard, "The Deep-Ocean Floor," *Scientific American*, September 1969

PAGES 34-35 (top)
After Bruce Heezen, Marie Tharp, and Maurice Ewing, *The Floors of the Oceans*, The Geological Society of America, Special Paper 65, 1959

PAGE 37 (bottom)
Collection of Edwin L. Hamilton

PAGE 38 (top)
John Shelton

PAGE 38 (bottom)
M. King Hubbert, "Strength of the Earth," *Bulletin of the American Association of Petroleum Geologists*, 29(1945):1630–1653

PAGE 40
National Maritime Museum, London

PAGE 41 (left)
Philip Alan Rosenberg/Pacific Stock

PAGE 41 (right)
Agatin Abbott, from *Volcanoes in the Sea*, by Gordon A. Macdonald, Agatin T. Abbott, and Frank L. Peterson, University of Hawaii Press, 1970

PAGES 46-47
Rob Wood—Stansbury, Ronsaville, Wood Inc.

PAGE 50
Tui De Roy

PAGE 54
Gudmundur Sigvaldason

PAGE 55 (top left)
Frans Lanting

PAGE 55 (top right)
Douglas Faulkner

PAGE 55 (bottom)
Robert Searle, Institute of Oceanographic Sciences, Wormley

PAGE 61 (top)
William F. Haxby, Lamont-Doherty Geological Observatory

PAGE 61 (bottom)
From *General Bathymetric Chart of the Oceans*, published by the Canadian Hydrographic Service, 1984

PAGE 64
William F. Haxby, Lamont-Doherty Geological Observatory

PAGE 70
Douglas Faulkner

PAGE 73 (left)
Clive Hambler

PAGE 73 (right)
Jenny Newing

PAGE 75 (top)
Christopher Kilburn

PAGE 79
After J. D. Milliman and K. O. Emery, "Sea Levels during the Past 35,000 Years," *Science* 162:1121-1123

PAGE 80
Paolo Pirazzoli

PAGE 81
Arthur L. Bloom

PAGE 83 (left)
Paolo Pirazzoli

PAGE 83 (right)
Erwin Christian

PAGE 86
Harold Simon/Tom Stack

PAGE 88
United States Geological Survey

PAGE 90
Frank Peterson, from *Volcanoes in the Sea*, 2nd edition, by Gordon A. Macdonald, Agatin T. Abbott, and Frank L. Peterson, University of Hawaii Press, 1983

PAGE 91 (left)
From *Fire under the Sea*, Lee Tepley, Moon-light Productions

PAGE 91 (right)
Woods Hole Oceanographic Institution

PAGE 94
After Charles D. Hollister, Morris F. Glenn, and Peter F. Lonsdale, "Morphology of Seamounts in the Western Pacific and Philippine Basin from Multi-beam Sonar Data," *Earth and Planetary Science Letters*, 41(1978):405-418

PAGE 97
United States Geological Survey

PAGE 98
From *General Bathymetric Chart of the Oceans*, published by the Canadian Hydrographic Service, 1984

PAGE 99
From *General Bathymetric Chart of the Oceans*, published by the Canadian Hydrographic Service, 1984

PAGE 102
David Muench

PAGE 103
Agatin Abbott, from *Volcanoes in the Sea*, by Gordon A. Macdonald, Agatin T. Abbot, and Frank L. Peterson, University of Hawaii Press, 1970

PAGE 104 (left and right)
D. W. Peterson, United States Geological Survey

PAGE 105
D. A. Swanson, United States Geological Survey

PAGE 107
Errol de Silva/Camera Hawaii

PAGE 108
John Béchervaise

PAGE 111
Tui De Roy

PAGE 112
Lauritz Sømme

PAGE 113
Werner Stoy/Camera Hawaii

PAGE 117
Australian Information Service

PAGE 119 (left)
Lucien Montaggioni

PAGE 119 (right)
Bureau de Recherches Geologiques et Minieres, Service Télédétection

PAGE 120
Erwin Christian

PAGE 121
Photo taken by astronaut Steven A. Hawley during mission 41-D of the space shuttle *Discovery*; courtesy of the Space Division, U.S. Air Force

PAGE 125
Mike James/The Photo Library, Australia

PAGE 128
United States Geological Survey

PAGE 131
Erwin Christian

PAGE 133
United States Geological Survey

PAGE 140 (top)
United States Geological Survey

PAGE 140 (bottom)
Erwin Christian

PAGE 146
Lucien Montaggioni

PAGE 150
Edwin L. Hamilton

PAGE 151
Lucien Montaggioni

PAGE 152
Lucien Montaggioni

PAGE 153
Paolo Pirazzoli

PAGE 154
From *General Bathymetric Chart of the Oceans*, published by the Canadian Hydrographic Survey, 1984

PAGE 159
United States Geological Survey

PAGE 161
Lucien Montaggioni

PAGE 163
Tui De Roy

PAGE 170
Jet Propulsion Laboratory

PAGE 173
A. Nur and Z. Ben-Avraham, "Speculations on Mountain Building and the Lost Pacifica Continent," *Journal of Physics*, Earth Supplement 26, 1978

PAGE 176
From *General Bathymetric Chart of the Oceans*, published by the *Canadian Hydrographic Service*, 1984

PAGE 177
Jean Francheteau, "The Oceanic Crust," *Scientific American*, September 1983

PAGE 180
Michael S. Marlow, David W. Scholl, Edwin C. Buffington, and Tau Rho Alpha, "Tectonic History of the Central Aleutian Arc," *Geological Society of America Bulletin*, 84:1555-1574, May 1973

PAGE 181
Travis Amos

PAGE 182
From *General Bathymetric Chart of the Oceans*, published by the Canadian Hydrographic Service, 1984

PAGE 188
From *General Bathymetric Chart of the Oceans*, published by the Canadian Hydrographic Service, 1984

PAGE 191
From *General Bathymetric Chart of the Oceans*, published by the Canadian Hydrographic Service, 1984

PAGE 193
After Amos Nur and Zvi Ben-Avraham, "Oceanic Plateaus, the Fragmentation of Continents, and Mountain Building," *Journal of Geophysical Research*, vol. 87, no. 15, pp. 3644–3661, May 10, 1982

PAGE 194
Flip Nicklin

PAGE 199
Field Museum of Natural History

PAGE 200
Peter Harrison

PAGE 202 (left)
Blacker-Wood Library of Zoology and Ornithology, McGill University, Montreal

PAGE 202 (right)
Frans Lanting

PAGE 204
National Maritime Museum, Greenwich

PAGE 206
National Archives

PAGE 207
The Bettman Archive

PAGE 208
India Office Library and Records

PAGE 209
Philip Alan Rosenberg/Pacific Stock

PAGE 210
Steven Stanley

PAGE 214
Richard W. Grigg

Index

Other Books in the Scientific American Library Series

POWERS OF TEN
by Philip and Phylis Morrison and the Office of Charles and Ray Eames

HUMAN DIVERSITY
by Richard Lewontin

THE DISCOVERY OF SUBATOMIC PARTICLES
by Steven Weinberg

THE SCIENCE OF MUSICAL SOUND
by John R. Pierce

FOSSILS AND THE HISTORY OF LIFE
by George Gaylord Simpson

THE SOLAR SYSTEM
by Roman Smoluchowski

ON SIZE AND LIFE
by Thomas A. McMahon and John Tyler Bonner

PERCEPTION
by Irvin Rock

CONSTRUCTING THE UNIVERSE
by David Layzer

THE SECOND LAW
by P. W. Atkins

THE LIVING CELL, VOLUMES I AND II
by Christian de Duve

MATHEMATICS AND OPTIMAL FORM
by Stefan Hildebrandt and Anthony Tromba

FIRE
by John W. Lyons

SUN AND EARTH
by Herbert Friedman

EINSTEIN'S LEGACY
by Julian Schwinger